Applied Mathematical Sciences | Volume 47

Applied Mathematical Sciences

1. John: **Partial Differential Equations**, 4th ed. (cloth)
2. Sirovich: **Techniques of Asymptotic Analysis.**
3. Hale: **Theory of Functional Differential Equations**, 2nd ed. (cloth)
4. Percus: **Combinatorial Methods.**
5. von Mises/Friedrichs: **Fluid Dynamics.**
6. Freiberger/Grenander: **A Short Course in Computational Probability and Statistics.**
7. Pipkin: **Lectures on Viscoelasticity Theory.**
8. Giacaglia: **Perturbation Methods in Non-Linear Systems.**
9. Friedrichs: **Spectral Theory of Operators in Hilbert Space.**
10. Stroud: **Numerical Quadrature and Solution of Ordinary Differential Equations.**
11. Wolovich: **Linear Multivariable Systems.**
12. Berkovitz: **Optimal Control Theory.**
13. Bluman/Cole: **Similarity Methods for Differential Equations.**
14. Yoshizawa: **Stability Theory and the Existence of Periodic Solutions and Almost Periodic Solutions.**
15. Braun: **Differential Equations and Their Applications**, 3rd ed. (cloth)
16. Lefschetz: **Applications of Algebraic Topology.**
17. Collatz/Wetterling: **Optimization Problems.**
18. Grenander: **Pattern Synthesis: Lectures in Pattern Theory, Vol I.**
19. Marsden/McCracken: **The Hopf Bifurcation and its Applications.**
20. Driver: **Ordinary and Delay Differential Equations.**
21. Courant/Friedrichs: **Supersonic Flow and Shock Waves.** (cloth)
22. Rouche/Habets/Laloy: **Stability Theory by Liapunov's Direct Method.**
23. Lamperti: **Stochastic Processes: A Survey of the Mathematical Theory.**
24. Grenander: **Pattern Analysis: Lectures in Pattern Theory, Vol. II.**
25. Davies: **Integral Transforms and Their Applications.**
26. Kushner/Clark: **Stochastic Approximation Methods for Constrained and Unconstrained Systems.**
27. de Boor: **A Practical Guide to Splines.**
28. Keilson: **Markov Chain Models—Rarity and Exponentiality.**
29. de Veubeke: **A Course in Elasticity.**
30. Sniatycki: **Geometric Quantization and Quantum Mechanics.**
31. Reid: **Sturmian Theory for Ordinary Differential Equations.**
32. Meis/Markowitz: **Numerical Solution of Partial Differential Equations.**
33. Grenander: **Regular Structures: Lectures in Pattern Theory, Vol. III.**
34. Kevorkian/Cole: **Perturbation Methods in Applied Mathematics.** (cloth)
35. Carr: **Applications of Centre Manifold Theory.**

(continued after Index)

Jack K. Hale
Luis T. Magalhães
Waldyr M. Oliva

An Introduction to Infinite Dimensional Dynamical Systems — Geometric Theory

With an Appendix by Krzysztof P. Rybakowski

With 17 Illustrations

Springer-Verlag
New York Berlin Heidelberg Tokyo

Jack K. Hale
Division of Applied Mathematics
Brown University
Providence, R.I. 02912
U.S.A.

Waldyr M. Oliva
Departmento de Matemática Aplicada
Instituto de Matemática e Estatistica
Universidade de São Paulo
São Paulo
Brasil

Luis T. Magalhães
Universidade Tecnica de Lisbõa
Lisbon
Portugal

Krzysztof P. Rybakowski
Technische Universität Berlin
Berlin
Federal Republic of Germany

AMS Subject Classifications: 34C35, 54H20

Library of Congress Cataloging in Publication Data
Hale, Jack K.
An introduction to infinite dimensional dynamical systems–geometric theory.
(Applied mathematical sciences; v. 47)
Bibliography: p.
Includes index.
1. Differentiable dynamical systems. I. Magalhães, Luis T. II. Oliva, Waldyr M. III. Title. IV. Series: Applied mathematical sciences (Springer-Verlag New York Inc.) ; v. 47.
Qa1.A647 vol. 47 [QA614.8] 510s [514′.7] 83-20043

Printed and bound by R.R. Donnelley & Sons, Harrisonburg, Virginia.
Printed in the United States of America.

9 8 7 6 5 4 3 2 1

ISBN 0-387-90931-1 Springer-Verlag New York Berlin Heidelberg Tokyo
ISBN 3-540-90931-1 Springer-Verlag Berlin Heidelberg New York Tokyo

Preface

The motivation for writing these notes came from a series of lectures of the third author on retarded functional differential equations at the Lefschetz Center for Dynamical Systems of the Division of Applied Mathematics at Brown University during the spring of 1982. Partial financial support was obtained from the Air Force Office of Scientific Research, AF-AFOSR 81-0198, National Science Foundation, MCS 79-05774-05, U. S. Army Research Office, DAAG-29-79-C-0161, Instituto Nacional de Investigação Cientifica, Portugal, Conselho Nacional de Desenvolvimento Cientifico e Tecnologico (CNPq), Proc. No. 40.3278/81, Fapesp and Reitoria da Universidade de São Paulo, Brasil.

The authors appreciate the efforts of Dorothy Libutti, Katherine MacDougall, and Nancy Gancz for the preparation of the manuscript.

Contents

1. Introduction

Many applications involve dynamical systems in non-locally compact infinite dimensional spaces; for example, dynamical systems generated by partial differential equations and delay differential or functional differential equations. Because of the complexities involved in doing detailed analysis in infinite dimensions, these systems often are approximated by finite dimensional dynamical systems generated by ordinary differential equations in $\mathbb{R}^n$ or on a n-dimensional manifold M. The global theory of such dynamical systems is then used to better understand the complete dynamics of the system and the way that the system behaves as physical parameters are varied.

Although some efforts are being made to extend the finite dimensional ideas to infinite dimensions, the global theory is still in its infancy. One reason the development has been so slow follows from the infinite dimensionality of the problems and the complexities that result from this fact. Probably a more important reason is a consequence of the fact that the persons who work in abstract finite dimensional dynamical systems are unaware of some of the "nice" systems that exist in infinite dimensions, systems whose basic structure may be amenable to a mathematical theory approaching the completeness that is known for finite dimensions and, at the same time, require new ideas of a fundamental character. On the other side, persons dealing with specific infinite dimensional problems in the applications often are not aware of the fact that detailed knowledge of the ideas in finite dimensional problems can be adapted to their problems.

The purpose of these notes is to outline an approach to the development of a theory of dynamical systems in infinite dimensions which is analogous to the theory of finite dimensions. The first problem is to find a class for which there is some hope of classification and yet general enough to include some interesting applications. Throughout the notes, the discussion centers around retarded functional differential equations although the techniques and several of the results apply to more general situations; in particular, to neutral functional differential equations, parabolic partial differential equations and some other types of partial differential equations.

In the introduction, we give an abstract formulation of a class of dynamical systems which occur frequently in the applications and state some of the basic properties and problems that should be studied.

Let X, Y, Z be Banach spaces (sometimes Banach manifolds) and let $\mathscr{X}^r = C^r(Y,Z)$, $r \geq 1$, be the set of functions from Y to Z which are bounded and uniformly continuous together with their derivatives up through order r. We impose the usual topology on $\mathscr{X}^r$. (In applications, other topologies may be needed; for example, the Whitney topology.) For each $f \in \mathscr{X}^r$, let $T_f(t): X \to X$, $t \geq 0$, be a strongly continuous semigroup of transformations on X. For each $x \in X$, we suppose $T_f(t)x$ is defined for $t \geq 0$ and is C^r in x.

We say a point $x_0 \in X$ has a <u>backward extension</u> if there is a $\varphi: (-\infty,0] \to X$ such that $\varphi(0) = x_0$ and $T_f(t)\varphi(\tau) = \varphi(t+\tau)$ for $0 \leq t \leq -\tau$, $\tau \leq 0$. If there is a backward extension φ through x_0, we define $T_f(t)x_0 = \varphi(t)$, $t \leq 0$. A set $M \subset X$ is <u>invariant</u> if, for each $x \in M$, $T_f(t)x$ is defined and belongs to M for $t \in (-\infty,\infty)$. The <u>orbit</u> $\gamma^+(x)$ through x is defined as $\gamma^+(x) = \cup_{t \geq 0} T_f(t)x$.

Let

$$A(f) = \{x \in X: T_f(t)x \text{ is defined and bounded for } t \in (-\infty,\infty)\}.$$

The set $A(f)$ contains much of the interesting information about the semigroup $T_f(t)$. In fact, it is very easy to verify the following result.

Proposition 1.1. *If $A(f)$ is compact, then $A(f)$ is maximal, compact, invariant. If, in addition, all orbits have compact closure, then $A(f)$ is a global attractor. Finally, if $T_f(t)$ is one-to-one on $A(f)$, then $T_f(t)$ is a continuous group on $A(f)$.*

The first difficulty in infinite dimensional systems is to decide how to compare two semigroups $T_f(t)$, $T_g(t)$. It seems to be almost impossible to make a comparison of any system on all or even an arbitrary bounded set of X. If $A(f)$ is compact, Proposition 1.1 indicates that all essential information is contained in $A(f)$. Thus, we define equivalence relatively to $A(f)$.

Definition 1.2. We say f is equivalent to g, $f \sim g$, if there is a homeomorphism $h: A(f) \to A(g)$ which preserves orbits and the sense of direction in time. We say f is *stable relative* to $A(f)$ or *A-stable* if there is a neighborhood V of f in $\mathscr{X}^r$ such that $g \in V$ implies $g \sim f$. We say f is a *bifurcation point* if f is not A-stable.

The basic problem is to discuss detailed properties of the set $A(f)$ and to determine how $A(f)$ and the structure of the flow on $A(f)$ change with f.

If $A(f)$ is not compact, very little is known at this time. It becomes important therefore to isolate a class of semigroups for which $A(f)$

is compact. If $T_f(t)$ is an α-contraction for $t > 0$ and $T_f(t)$ is compact dissipative, then it can be proved that $A(f)$ is compact. We define in a later section an α-contraction, but it is sufficient at this time to note that a special case which is very important in the applications is

$$T_f(t) = S_f(t) + U_f(t), \quad t \geq 0,$$

where $S_f(t)$ is a strict contraction for $t > 0$ and $U_f(t)$ is completely continuous for $t \geq 0$. Compact dissipative means there is a bounded set B in X such that, for any compact set K in X, there is a $t_0 = t_0(K,B)$ such that $T_f(t)\, K \subset B$, $t \geq t_0$.

If $T_f(t)$ is completely continuous for $t \geq r$ for some $r > 0$, then it can be shown that $A(f)$ is compact if $T_f(t)$ is point dissipative, that is, each orbit eventually enters into a bounded set and remains.

Before proceeding further, we give two examples of semigroups which can be used as models to illustrate several of the ideas.

Suppose $u \in \mathbb{R}^k$, $x \in \mathbb{R}^n$, Ω is a bounded, open set in $\mathbb{R}^n$ with smooth boundary, D is a $k \times k$ constant diagonal, positive matrix, Δ is the Laplacian operator, and consider the equation

$$\begin{aligned} u_t - D\Delta u &= f(x,u,\text{grad } u) \quad \text{in } \Omega \\ u &= 0 \quad \text{on } \partial\Omega. \end{aligned}$$

Other boundary conditions could also be used. Let $W = W_0^{1,2}(\Omega) \cap W^{2,2}(\Omega)$ be the domain of $-\Delta$ and let $X = W^\alpha$, $0 < \alpha < 1$, be the domain of the fractional power $(-\Delta)^\alpha$ of $-\Delta$ with the graph norm. Under appropriate conditions, this equation generates a strongly continuous semigroup $T_f(t)$

on X which is compact for $t > 0$. In this case $\mathscr{X}^r = C^r(\Omega \times \mathbb{R}^k \times \mathbb{R}^{kn}, \mathbb{R}^k)$. If f is independent of x, then $\mathscr{X}^r = C^r(\mathbb{R}^k \times \mathbb{R}^{kn}, \mathbb{R}^k)$. If f depends only on u, then $\mathscr{X}^r = C^r(\mathbb{R}^k, \mathbb{R}^k)$. In each of these cases, the theory will be different.

As another example, suppose $r > 0$, $C = C([-r,0], \mathbb{R}^n)$, $\mathscr{X}^r = C^r(C, \mathbb{R}^n)$, $r \geq 1$, and consider the RFDE,

$$\dot{x}(t) = f(x_t)$$

where, for each fixed t, x_t designates the restriction of a function x as $x_t(\theta) = x(t+\theta)$, $-r \leq \theta \leq 0$. For any $\varphi \in C$, let $x(\varphi)(t)$, $t \geq 0$, designate the solution with $x_0(\varphi) = \varphi$ and define $T_f(t)\varphi = x_t(\varphi)$. If this function is defined for $t \geq 0$, then $T_f(t): C \to C$ is a strongly continuous semigroup and $T_f(t)$ is completely continuous for $t \geq r$ if it takes bounded sets to bounded sets.

For differential difference equations

$$\dot{x}(t) = f(x(t), x(t-r))$$
$$\dot{x}(t) = f(x(t-r))$$

the space $\mathscr{X}^r$ is respectively, $C^r(\mathbb{R}^n \times \mathbb{R}^n, \mathbb{R}^n)$, $C^r(\mathbb{R}^n, \mathbb{R}^n)$.

The abstract dynamical system above also include some neutral functional differential equations and other classes of partial differential equations.

Some basic questions that should be discussed are the following:

Q.1. Is $T_f(t)$ one-to-one on $A(f)$ generically in f?

Q.2. If f is A-stable, is $T_f(t)$ one-to-one on $A(f)$?

Q.3. *When is* $A(f)$ *a manifold or a finite union of manifolds?*

Q.4. *Can* $A(f)$ *be imbedded in a finite dimensional manifold generically in* f?

Q.5. *For each* $x \in A(f)$, *is* $T_f(t)x$ *continuously differentiable in* $t \in \mathbb{R}$?

Q.6. *Are Kupka-Smale semigroups generic?*

Q.7. *Are Morse-Smale systems open and A-stable?*

Notice that all questions are posed for $A(f)$.

In these notes, we are going to discuss in detail how one can obtain a geometric theory for retarded functional differential equations and we attempt to answer some of the questions above. Throughout the notes, we will point out when the techniques and results are applicable to the more general abstract framework. We have attempted to give a unified exposition of some of the fundamental results in this subject, always making the presentation as self-contained as possible. Some parts of the notes are also devoted to speculations on the directions for future research.

2. Retarded Functional Differential Equations on Manifolds

Let M be a separable C^∞ finite dimensional connected manifold, I the closed interval $[-r,0]$, $r > 0$, and $C^0(I,M)$ the totality of continuous maps φ of I into M. Let TM be the tangent bundle of M and τ_M: $TM \to M$ its C^∞-canonical projection. Assume there is given on M a complete Riemannian structure (it exists because M is separable) with δ_M the associated complete metric. This metric on M induces an admissible metric on $C^0(I,M)$ by

$$\delta(\varphi,\overline{\varphi}) = \sup\{\delta_M(\varphi(\theta),\overline{\varphi}(\theta)): \theta \in I\}.$$

The space $C^0(I,M)$ is complete and separable, because M is complete and separable. The function space $C^0(I,M)$ is a C^∞-manifold modeled on a separable Banach space. If M is imbedded as a closed submanifold of an Euclidean space V, then $C^0(I,M)$ is a closed C^∞-submanifold of the Banach space $C^0(I,V)$.

If ρ: $C^0(I,M) \to M$ is the evaluation map, $\rho(\varphi) = \varphi(0)$, then ρ is C^∞ and, for each $a \in M$, $\rho^{-1}(a)$ is a closed submanifold of $C^0(I,M)$ of co-dimension $n = \dim M$. A <u>retarded functional differential equation (RFDE) on</u> M is a continuous function F: $C^0(I,M) \to TM$, such that $\tau_M F = \rho$. Roughly speaking, an RFDE on M is a function mapping each continuous path φ lying on M, $\varphi \in C^0(I,M)$, into a vector tangent to M at the point $\varphi(0)$. The notation RFDE(F) is used as short for "retarded functional differential equation F". Nonautonomous RFDE's on manifolds could be similarly defined, but we restrict the definition to the autonomous case as these are the only equations discussed in the present notes.

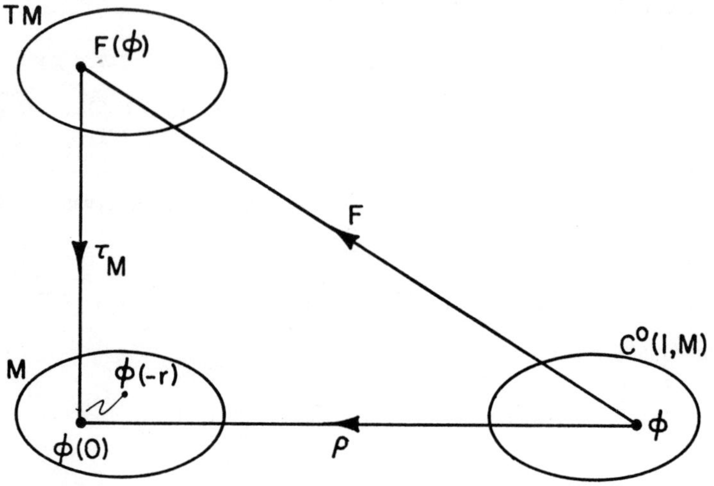

Figure 2.1

Given a function x of a real variable and with values in the manifold M, we denote $x_t(\theta) = x(t+\theta)$, $\theta \in I$, whenever the right-hand side is defined for all $\theta \in I$. A <u>solution of an RFDE(F) on M</u> with initial condition $\varphi \in C^0(I,M)$ at t_0 is a continuous function $x(t)$ with values on M and defined on $t_0 - r \leq t < t_0 + A$, for some $0 < A < \infty$, such that:

(i) $x_{t_0} = \varphi$,

(ii) $x(t)$ is a C^1-function of $t \in [t_0, t_0+A)$,

(iii) $(x(t),(d/dt)x(t)) = F(x_t)$, $t \in [t_0, t_0+A)$,

where $(x(t),(d/dt)(x(t))$ denotes the tangent vector to the curve $x(t)$ at the point t.

One can write locally

$$(x(t),\dot{x}(t)) = F(x_t) = (x(t),f(x_t))$$

or simply

$$\dot{x}(t) = f(x_t),$$

for an appropriate function f.

An existence and uniqueness theorem for initial value problems can be established with basis on the corresponding result for $M = \mathbb{R}^n$. A function G between two Banach manifolds is said to be locally Lipschitzian at a certain point φ of its domain, if there exist coordinate neighborhoods of φ and of $G(\varphi)$, in the domain and in the range of G, respectively, and the representation of G defined through the associated charts is Lipschitz, as a mapping between subsets of Banach spaces.

Theorem 2.1. If F is an RFDE on M which is locally Lipschitzian, then for each $\varphi \in C^0(I,M)$, $t_0 \in \mathbb{R}$, there exists a unique solution $x(t)$ of F with initial condition $x_{t_0} = \varphi$.

Proof: By Whitney's embedding theorem, M can be considered as a submanifold of $\mathbb{R}^N$ for an appropriate integer N. Accordingly, TM can be considered a submanifold of $\mathbb{R}^N \times \mathbb{R}^N$. We will construct an extension $\overline{F}$ of F which defines an RFDE $\overline{F}: C^0(I, \mathbb{R}^N) \to \mathbb{R}^N \times \mathbb{R}^N$ such that $\overline{F}(\varphi) = F(\varphi)$ if $\varphi \in C^0(I,M)$ and $\overline{F}(\varphi) = 0$ outside a certain neighborhood of $C^0(I,M)$ in $C^0(I, \mathbb{R}^N)$. Let U be a tubular neighborhood of M in $\mathbb{R}^N$ and α the C^∞ projection. Let W be the open set $W = \{\varphi \in C^0(I, \mathbb{R}^N) \mid \varphi(I) \subset U\}$. Define $F_1: W \to \mathbb{R}$ by $F_1(\varphi) = 1 - \int_{-r}^{0} |\alpha(\varphi(s)) - \varphi(s)|^2 ds$ where $|\cdot|$ is the Euclidean norm in $\mathbb{R}^N$. Then $F_1(\varphi) = 1$ if and only if $\varphi \in C^0(I,M)$ and $F_1(\varphi) < 1$ if $\varphi \notin C^0(I,M)$. For every $0 < \varepsilon < 1$ let $W_\varepsilon = F_1^{-1}([1-\varepsilon,\infty))$. Fix some $0 < \varepsilon < 1$ and take a C^∞ $\psi: \mathbb{R} \to \mathbb{R}$ satisfying $\psi(t) = 1$ for

$t \geq 1$ and $\psi(t) = 0$ for $t \leq 1 - \varepsilon/2$. Define $F_2\colon C^0(I, \mathbb{R}^N) \to \mathbb{R}$ as $F_2(\varphi) = \psi(F_1(\varphi))$ if $\varphi \in W_\varepsilon$ and $F_2(\varphi) = 0$ if $\varphi \notin W_\varepsilon$. Then F_2 is a C^∞-function and satisfies $F_2(\varphi) \leq 1$ for all φ and $F_2(\varphi) = 1$ if and only if $\varphi \in C^0(I,M)$. Finally define $\overline{F}$ as $\overline{F}(\varphi) = 0$ when $\varphi \notin W$ and $\overline{F}(\varphi) = F_2(\varphi)F(\alpha\circ\varphi)$ when $\varphi \in W$. The standard results on existence and uniqueness of solutions of FDE on $\mathbb{R}^N$ can be applied to finish the proof of the theorem.

Using the ideas in the proof of Theorem 2.1, it is possible to establish, for RFDE's on manifolds, results on continuation of solutions to maximal intervals of existence, and on continuous dependence relative to changes in initial data and in the RFDE, which are analogous to the corresponding results in $\mathbb{R}^n$.

Given a locally Lipschitzian RFDE(F) on M, its maximal solution $x(t)$, satisfying the initial condition φ at t_0 is sometimes denoted by $x(t;t_0,\varphi,F)$, and x_t is denoted by $x_t(t_0,\varphi,F)$. The arguments φ and F will be dropped whenever confusion may not arise, and t_0 will be dropped if $t_0 = 0$.

The <u>solution map</u> or <u>semiflow</u> of an RFDE(F) is defined by $\Phi(t,\varphi,F) = x_t(\varphi,F)$, whenever the right-hand side makes sense. It will be written as $\Phi(t,\varphi)$, whenever confusion is not possible. The notation $\Phi_t\varphi = \Phi(t,\varphi)$ is also used.

The following theorem gives some important properties of the semiflow Φ. For the statement of differentiability properties of Φ, it is convenient to introduce the notation $\mathcal{X}^k = \mathcal{X}^k(I,M)$, $k \geq 1$, for the Banach space of all C^k-RFDE's defined on the manifold M, which are bounded and have bounded derivatives up to order k, taken with the C^k-uniform norm.

Theorem 2.2. *If* F *is an RFDE on* M *in* X^k, $k \geq 1$ *then the family of mappings* $\{\Phi_t, 0 < t < \infty\}$ *is a strongly continuous semigroup of operators on* $C^0 = C^0(I,M)$ *ie.*,

(i) Φ_0 *is the identity on* C^0,

(ii) $\Phi_{t+s} = \Phi_t\Phi_s$, $t,s \geq 0$, $t + s < T$

(iii) *the map from* $[0,\infty) \times C^0$ *into* C^0 *given by* $(t,\varphi) \to \Phi_t(\varphi)$ *is continuous.*

Furthermore, the solution map, as a function $\Phi\colon [0,\infty) \times C^0 \times \mathscr{X}^k \to C^0$, $k \geq 1$, *has the following regularity properties:*

1) Φ *is continuous on* $[0,\infty) \times C^0 \times \mathscr{X}^k$,

2) *for each fixed* $t \geq 0$, *the map* $\Phi(t,\cdot,\cdot)\colon C^0 \times \mathscr{X}^k \to C^0$ *is* C^k,

3) *the map* $\Phi\colon (sr,\infty) \times C^0 \times \mathscr{X}^k \to C^0$ *is* C^s, *for all* $0 \leq s \leq k$.

Proof: Using the ideas in the proof of Theorem 2.1, these properties can be reduced to the analogous properties for RFDE's in $\mathbb{R}^n$.

The map $\Phi_t\colon C^0 \to C^0$ needs not be one-to-one, but, if there exists $\varphi,\psi \in C^0$ and $t,s \geq 0$ such that $\Phi_t\varphi = \Phi_s\psi$, then $\Phi_{t+\sigma}(\psi) = \Phi_{s+\sigma}(\psi)$ for all $\sigma \geq 0$ for which these terms are defined.

The following property of the solution map is also useful.

Theorem 2.3. *If* F *is an RFDE,* $F \in \mathscr{X}^1$ *and the corresponding solution map* $\Phi_t\colon C^0([-r,0],M) \to C^0([-r,0],M)$ *is uniformly bounded on compact subsets of* $[0,\infty)$, *then, for* $t \geq r$, Φ_t *is a compact map, i.e., it maps bounded sets of* C^0 *into relatively compact subsets of* C^0.

Proof: Again, this property can be reduced to the analogous property for

FDE's in $\mathbb{R}^n$. Actually, the proof is an application of the Ascoli-Arzela theorem.

A consequence of this result is that, for an RFDE(F) satisfying the hypothesis in the theorem with $r > 0$, Φ_t can never be a homeomorphism because the unit ball in $C([-r,0], \mathbb{R}^n)$ is not compact. The hypothesis of Theorem 2.3 are satisfied if $F \in \mathcal{Q}^k$ and M is compact.

The double tangent space, T^2M, of the manifold M, admits a canonical involution $w: T^2M \to T^2M$, w^2 equal to the identity on T^2M, and w is a C^∞-diffeomorphism on T^2M which satisfies $\tau_{TM} \cdot w = T\tau_M$ and $T\tau_M \cdot w = \tau_{TM}$, where $\tau_M: TM \to M$ and $\tau_{TM}: T^2M \to TM$ are the corresponding canonical projections. If F is a C^k RFDE on a manifold M, $k \geq 1$, and TF is its derivative, it follows that $w \cdot TF$ is a C^{k-1} RFDE on TM, which is called the <u>first variational equation of</u> F. The map w is norm preserving on T^2M, and the solution map Ψ_t, of $w \cdot TF$ is the derivative of the solution map, Φ_t, of F, i.e., $\Psi_t = T\Phi_t$.

3. Examples of Retarded Functional Differential Equations on Manifolds

3.1. RFDE's on $\mathbb{R}^n$.

Autonomous retarded functional differential equations on $\mathbb{R}^n$ are usually defined as equations of the form

$$\dot{x}(t) = f(x_t)$$

where f maps $C^0(I, \mathbb{R}^n)$ into $\mathbb{R}^n$. Taking $M = \mathbb{R}^n$ and identifying TM with $\mathbb{R}^n \times \mathbb{R}^n$, one can define the function $F: C^0(I,M) \to TM$ such that $F(\varphi) = (\varphi(0), f(\varphi))$. If f is continuous, then F is an RFDE on $M = \mathbb{R}^n$ which can be identified with the above equation.

3.2. Ordinary Differential Equations as RFDE's.

Any continuous vector field X on a manifold M defines an RFDE on M by $F = X\rho$ where $\rho: C^0 \to M$ is, as before, the evaluation map $\rho(\varphi) = \varphi(0)$.

3.3. Ordinary Differential Equations on $C^0(I,M)$.

Any continuous vector field Z on $C^0 = C^0(I,M)$ for $I = [-r,0]$, $r > 0$ and M a manifold, defines an RFDE on M by $F = T\rho \circ Z$, where $T\rho$ denotes the derivative of the evaluation map ρ.

3.4. Products of Real Functions on $C^0(I,M)$ by RFDE's on M.

If $g: C^0 \to \mathbb{R}$ is continuous and F is an RFDE on M, then the map $G: C^0 \to TM$ given by $G(\varphi) = g(\varphi) \cdot F(\varphi)$ is also an RFDE on M.

3.5. RFDE's on TM.

Retarded functional differential equations on TM are continuous maps $\overline{F}: C^0(I,TM) \to T^2M$ satisfying $\tau_{TM} \cdot \overline{F} = T\rho$.

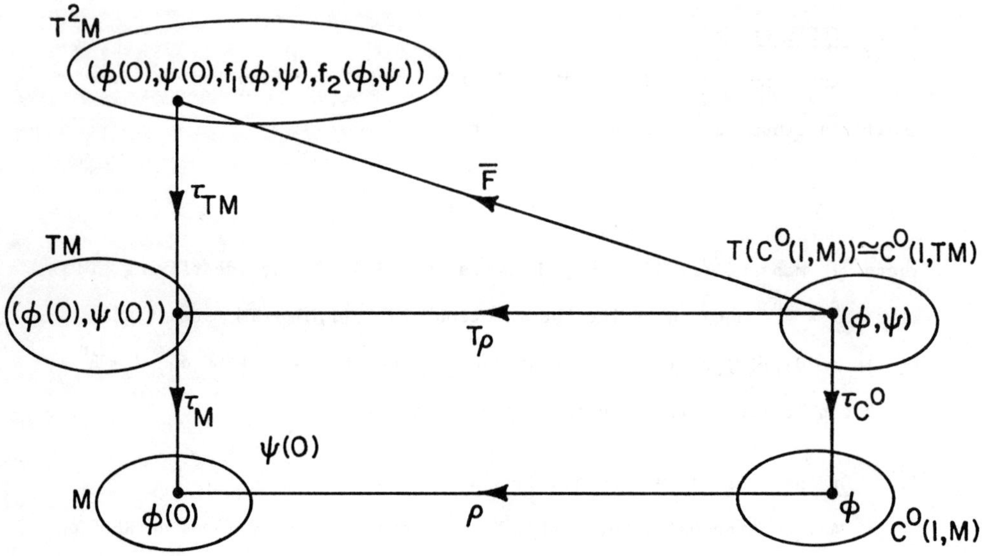

Figure 3.1

Recall that one can write locally $\overline{F}(\varphi,\psi) = (\varphi(0),\psi(0),f_1(\varphi,\psi), f_2(\varphi,\psi))$. Consequently, for the solutions $(x(t),y(t))$ on TM we have

$$\dot{x}(t) = f_1(x_t,y_t)$$
$$\dot{y}(t) = f_2(x_t,y_t).$$

Given a C^1 RFDE(F), its first variational equation $\overline{F}$ is a special case of an RFDE on TM. Denoting locally, $F(\varphi) = (\varphi(0),f(\varphi))$, we have $\overline{F}(\varphi,\psi) = (\varphi(0),\psi(0),f(\varphi),df(\varphi)\psi)$, where df denotes the derivative of f. The solutions $(x(t),y(t))$ on TM, of the first variational equation $\overline{F}$ must satisfy

$$\dot{x}(t) = f(x_t)$$

$$\dot{y}(t) = df(x_t)y_t.$$

3.6. Second order RFDE's on M.

Another special case of RFDE's on TM is associated with second order RFDE's on M.

Let $\overline{F}: C^0(I,TM) \to T^2M$ a continuous function such that, locally,

$$\overline{F}(\varphi,\psi) = (\varphi(0),\psi(0),\psi(0),f(\varphi,\psi)).$$

The solutions $(x(t),y(t))$ of the RFDE($\overline{F}$) on TM satisfy

$$\dot{x}(t) = y(t)$$

$$\dot{y}(t) = f(x_t,y_t)$$

or

$$\ddot{x}(t) = f(x_t,\dot{x}_t)$$

where $x(t)$ assumes values in M. We are therefore justified in calling second order RFDE's on M to the functions $\overline{F}: C^0(I,TM) \to T^2M$ of the form described above.

3.7. Differential Delay Equations on M.

Let $g: M \times M \to TM$ be a continuous function such that $\tau_M \cdot g = \pi_1$ is the first projection of $M \times M$ upon M, and let $d: C^0(I,M) \to M \times M$ be such that $d(\varphi) = (\varphi(0),\varphi(-r))$.

The function $F = g \cdot d$ is an RFDE on M, and for its solutions $x(t)$ one can write, locally,

$$(x(t),\dot{x}(t)) = g(x(t),x(t-r)) = (x(t),\bar{g}(x(t),x(t-r)))$$

or simply

$$\dot{x}(t) = \bar{g}(x(t),x(t-r)).$$

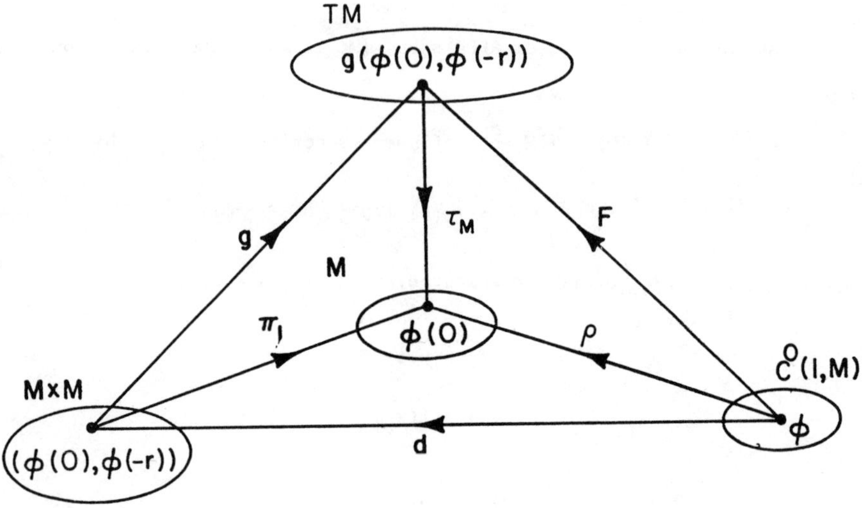

Figure 3.2

3.8. RFDE's on Imbedded Submanifolds of $\mathbb{R}^n$.

Let S be an imbedded submanifold of $\mathbb{R}^n$ which is positively invariant under the RFDE on $\mathbb{R}^n$ given by

$$\dot{x}(t) = f(x_t),$$

i.e., solutions with initial condition φ at $t = 0$ such that $\varphi(0) \in S$, assume values in S for all $t \geq 0$ in their interval of existence.

The function $F: C^0(I,S) \to TS$ such that $F(\varphi) = (\varphi(0),f(\varphi))$ is an RFDE on S.

3.9. <u>An RFDE on S^2.</u>

Let us consider the system of differential delay equations on $\mathbb{R}^3$

$$\dot{x}(t) = -x(t-1)y(t) - z(t)$$
$$\dot{y}(t) = x(t-1)x(t) - z(t)$$
$$\dot{z}(t) = x(t) + y(t).$$

Its solutions satisfy $x\dot{x} + y\dot{y} + z\dot{z} = 0$, or $x^2 + y^2 + z^2 = \text{constant}$, $t \geq 0$. Consequently, if $\varphi \in C^0([-1,0]; \mathbb{R}^3)$ and $\varphi(0) \in S^2 = \{(x,y,z) \in \mathbb{R}^3: x^2 + y^2 + z^2 = 1\}$, S^2 is positively invariant and, therefore, the given system induces an RFDE on S^2 by the construction given in the preceding example.

3.10. <u>RFDE's on S^1.</u>

a) The set $S^1 = \{(x,y) \in \mathbb{R}^2: x^2 + y^2 = 1\}$ is positively invariant under the system

$$\dot{x}(t) = -y(t)(1-x(t))x(t-1)$$
$$\dot{y}(t) = x(t)(1-x(t))x(t-1).$$

Consequently, this system induces an RFDE on S^1 by the procedure outlined in 3.8.

b) The equation defined on $\mathbb{R}$ by

$$\dot{x}(t) = k \sin (x(t)-x(t-1))$$

can be considered as an RFDE(F) on S^1.

In fact, the map $\gamma: [0,2\pi) \to S^1$ such that $\gamma(\beta) = (\cos \beta, \sin \beta)$ is one-to-one and onto and we can define $F: C^0([-1,0], S^1) \to TS^1$ by

$$F(\varphi) = (\varphi(0),\ k\sin(\varphi(0) - \varphi(-1))u_{\varphi(0)})$$

where $u_{\varphi(0)}$ is a unit vector tangent to S^1 at $\varphi(0)$ and using the usual identification $TS^1 = S^1 \times \mathbb{R}$.

c) The equation defined on $\mathbb{R}$ by

$$\dot{x}(t) = \frac{\pi}{2}(1-\cos x(t)) + \frac{\pi}{2}(1-\cos x(t-1))$$

is another example of an equation that can be considered as an RFDE on S^1 by the same procedure used in b).

3.11. <u>A Second Order Equation on S^1.</u>

The second order scalar equation

$$\ddot{x}(t) = A\dot{x}(t) + B\sin x(t-r)$$

can be written as a system

$$\dot{x}(t) = y(t)$$
$$\dot{y}(t) = Ay(t) + B\sin x(t-r)$$

where $A,B \in \mathbb{R}$. This system defines a second order RFDE on S^1 given by map $F: C^0(I,TS^1) \to T^2S^1$ such that

$$F(\varphi,\psi) = (\varphi(0),\psi(0),\psi(0),\ A\psi(0) + B\sin\varphi(-r)).$$

As a matter of fact, this equation is an RFDE on the cylinder $S^1 \times \mathbb{R} = TS^1$.

This equation has been studied in connection with the circumutation of plants and is sometimes called the sunflower equation.

3.12. The Levin-Nohel Equation on S^1.

Let $G: \mathbb{R} \to \mathbb{R}$ and $a: [0,r] \to \mathbb{R}$ be C^1 functions and denote by g the derivative of G. The scalar equation

$$\dot{x}(t) = -\int_{-r}^{0} a(-\theta)g(x(t+\theta))d\theta$$

is known as the Levin-Nohel equation. It has been studied in connection with nuclear reactor dynamics.

A special case of this equation is obtained with $G(x) = 1-\cos x$. This equation can be considered as an RFDE on S^1 by the same procedure as used for the example in 3.10 b), c).

3.13. Equations Obtained by Compactification.

In the study of polynomial vector fields in the plane $\mathbb{R}^2$, Poincaré used a compactification of $\mathbb{R}^2$ given by a central projection of $\mathbb{R}^2$ into a unit sphere S^2 tangent to the plane $\mathbb{R}^2$ at the origin, when these manifolds are considered as imbedded in $\mathbb{R}^3$. This compactification procedure can be extended to construct delay equations on spheres from polynomial equations on $\mathbb{R}^n$, $n \geq 1$.

For the purposes of illustration, let us consider any of the following differential equations on $\mathbb{R}$:

$$\dot{x}(t) = P(x(t))$$

or

$$\dot{x}(t) = P(x(t-1)) \tag{3.1}$$

or

$$\dot{x}(t) = P(x(t),x(t-1)),$$

where P is a polynomial of degree p. One can consider the central projection at the line $\{(x,1): x \in \mathbb{R}\}$ into the circle $S^1 = \{(y_1,y_2) \in \mathbb{R}^2: y_1^2 + y_2^2 = 1\}$, given by $(y_1,y_2) = \pm(x,1)/\Delta(x)$, with $\Delta(x) = (1+x^2)^{1/2}$.

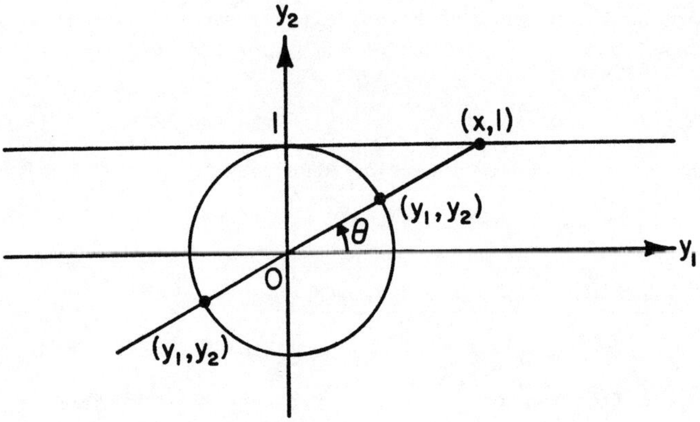

Figure 3.3

For $y_2 \neq 0$, we have $x = y_1/y_2$ and therefore

$$(y_2\dot{y}_1 - y_1\dot{y}_2)/y_2^2 = \dot{x} = P(*)\cdot[F]$$

$$(y_1\dot{y}_1 + y_2\dot{y}_2) = 0,$$

where $P(*)$ denotes the right-hand side of the particular equation in (3.1) which is being considered and [F] denotes an appropriate multiplication factor to be chosen according to the application envisaged. Solving the last system of equations for $\dot{y}_1$ and $\dot{y}_2$, one obtains

$$\dot{y}_1 = y_2^2P(*)y_2[F]$$
$$\dot{y}_2 = y_1y_2P(*)y_2[F].$$

The particular case

$$P(*) = P(x(t-1)) = -kx(t-1),$$

gives, under the above central projection,

$$P(y_1(t-1)/y_2(t-1)) = -ky_1(t-1)/y_2(t-1).$$

Choosing for multiplicative factor $[F] = y_2(t-1)/y_2(t)$ one gets

$$\dot{y}_1(t) = -ky_2^2(t)y_1(t-1)$$

$$\dot{y}_2(t) = ky_1(t)y_2(t)y_1(t-1).$$

This system can be considered as an RFDE on S^1 by a procedure similar to the one used for the examples, 3.10 b,c. In terms of the angle coordinate (see Fig. 3.3), the equation can also be written as

$$\dot{\theta}(t) = k \sin \theta(t) \cdot \cos \theta(t-1). \qquad (3.2)$$

A different choice for the multiplicative factor $[F]$ could be used to obtain a different equation on S^1. For instance $[F] = y_2(t-1)$, would lead to the RFDE on S^1 given by

$$\dot{y}_1(t) = -ky_2(t)y_1(t-1)$$

$$\dot{y}_2(t) = ky_1(t)y_1(t-1)$$

or, in the polar coordinate θ,

$$\dot{\theta}(t) = k \cos \theta(t-1). \qquad (3.3)$$

The multiplicative factor $[F]$ is to be chosen according to the application at hand. If, for instance, the study at infinity in the original coordinate is desired, it is convenient to choose [F] so that the points on S^1 corresponding to $+\infty$ and $-\infty$ in the original coordinate, $\theta = 0$ and $\theta = \pi$, respectively, be invariant under the induced RFDE on S^1. It can be seen from (3.2) and (3.3), that this is the case for the first factor [F] used above, but not for the second. Actually, in the original coordinates, the equation obtained with the factor

$$[F] = y_2(t-1)/y_2(t) = [(1+x^2(t))/(1+x^2(t-1))]^{1/2}$$

is

$$\dot{x}(t) = -kx(t-1)[1+x^2(t-1)]^{1/2},$$

and the equation obtained with the factor

$$[F] = y_2(t-1) = (1+x^2(t-1))^{-1/2}$$

is

$$\dot{x}(t) = -kx(t-1)[1+x^2(t-1)]^{-1/2}.$$

A similar Poincaré compactification can be used for higher dimensions. In particular, given a delay equation in $\mathbb{R}^2$

$$\dot{x}(t) = Ax(t-1),$$

where A is a 2×2 real matrix, the Poincare compactification of $\mathbb{R}^2$ into the unit sphere S^2 tangent to $\mathbb{R}^2$ at the origin (with both manifolds considered as imbedded in $\mathbb{R}^3$), leads to an RFDE on the unit sphere S^2.

3.14. <u>The Linear Variational Equation of a C^1 RFDE(F) at a Solution $p = p(t)$</u>.

Let F be a C^1 RFDE on a manifold M and $p(t)$ a fixed solution of F, $t \in [\sigma-r,A)$. The linear variational equation of F at $p(t)$ is the restriction of the first variational equation of F to the set $R(F,p) = \tau^{-1}_{C^0}\{p_t \mid t \in [\sigma,A)\}$, which is a subset of $TC^0(I,M)$, positively invariant under the flow of the first variational equation of F. A "solution" of the linear variational equation at $p(t)$ is a solution of the first variational equation of F with initial condition in $R(F;p)$. That is, a solution locally defined by $(p(t),y(t))$ satisfying $\dot{y}(t) = df(p_t)y_t$ for the appropriate f (see Example 3.5). The equation $\dot{y}(t) = L(t)\cdot y_t$, $t \in [\sigma,A)$, $L(t) = df(p_t)$, is the local representation of the linear variational equation at $p(t)$.

4. Generic Properties. The Theorem of Kupka-Smale

The aim of the generic theory of differential equations is to study qualitative properties which are typical of the class of equations considered, in the sense that they hold for all equations defined by functions of a residual set of the function space being considered. More precisely, if X is a complete metric space, then a property $\mathcal{P}$ on the elements $x \in X$ is said to be generic if there is a residual set $Y \subset X$ such that each element of Y has property $\mathcal{P}$. Recall that a residual set is a countable intersection of open dense sets. As for ordinary differential equations, the constant and the periodic solutions, and their stable and unstable manifolds, play an important role in the generic theory of RFDE's.

Given an RFDE(F) on a manifold M, we say that a constant function $\varphi \in C^0(I,M)$ is a critical point or an equilibrium point of F, if the solution of F with initial data φ is constant, i.e., $F(\varphi) = 0$. A critical point φ of F is said to be nondegenerate if zero is not a characteristic value of the linear variational equation of F at φ; φ is said to be hyperbolic if there is no characteristic value of the linear variational equation having real part equal to zero. Locally, the RFDE(F) on M can be identified with an equation on euclidean space, and there exist manifolds $W^s_{loc}(\varphi)$ and $W^u_{loc}(\varphi)$ - the local stable manifold and local unstable manifold of F at φ - which have the property that, for some $\varepsilon_0 > 0$ and all balls $B_\varepsilon(\varphi) = \{\psi \in C^0: \text{dist}(\varphi,\phi) \leq \varepsilon\}$, $0 < \varepsilon < \varepsilon_0$ they consist of the points in orbits of F which stay in $B_\varepsilon(\varphi)$ for all $t \geq 0$ or $t \leq 0$, respectively. The manifolds $W^s_{loc}(\varphi)$ and $W^u_{loc}(\varphi)$ are "tangent" at φ to linear manifolds S and U which decompose, as a direct sum, the phase space of the linear variational equation of F. The dimension of $W^u_{loc}(\varphi)$ and U is finite. The solutions x(t) of F with initial data in W^u_{loc} are defined for all $-\infty < t < 0$, and the union of the complete orbits having initial data in $W^u_{loc}(\varphi)$ defines in some cases a manifold $W^u(\varphi)$ called the <u>global unstable manifold of</u> F <u>at</u> φ. The flow of F on the finite dimensional manifold $W^u_{loc}(\varphi)$ can be associated with an ordinary differential equation.

The concepts of nondegeneracy and hyperbolicity can also be defined for periodic solutions of an RFDE(F). If p(t) is a nonconstant ω-periodic solution of F, by compactness of the intervals $[t_0, t_0+T]$, the RFDE(F) on M can be identified, locally around p(t), with an equation on Euclidean space $\mathbb{R}^n$. One can then consider the linear variational equation relative to p(t). This equation is a linear periodic system of period ω, having $\dot{p}(t)$ as one of its solutions. It follows that $\mu = 1$ is a characteristic multiplier of the linear variational equation relative to p(t). We say the <u>periodic orbit</u> $\Gamma = \{p(t),\ t \in \mathbb{R}\}$ is <u>nondegenerate</u> if the characteristic multiplier $\mu = 1$ is simple and we say <u>the periodic orbit</u> Γ <u>is hyperbolic</u> if it is nondegenerate and $\mu = 1$ is the only characteristic multiplier with $|\mu| = 1$.

The theorem of Kupka-Smale for ordinary differential equations, perhaps the most basic result of generic theory, asserts that the property that all critical points and periodic orbits are hyperbolic and the stable and unstable manifolds intersect transversally is generic in the class of all ordinary differential equations $\dot{x} = f(x)$, $x \in \mathbb{R}^n$ or $x \in M$ (M a compact submanifold of $\mathbb{R}^n$) for which f is smooth in an adequate topology. The complete proof of the Kupka-Smale theorem for RFDE's is not presently available, but some results in this direction are known.

The first generic results for RFDE's were established for equations defined on a compact manifold M, proving that the sets G_0^k and G_1^k of all RFDE's in $\mathscr{X}^k(I,M)$ which have all critical points nondegenerate and hyperbolic, respectively, are open and dense in $\mathscr{X}^k(I,M)$, $k \geq 1$, and the sets $G_{3/2}^k(T)$ and $G_2^k(T)$ of all RFDE's in $\mathscr{X}^k(I,M)$ for which all nonconstant periodic solutions with period in $(0,T]$ are nondegenerate and hyperbolic, respectively, are open in $\mathscr{X}^k(I,M)$, $k \geq 1$. For the case of RFDE's on $\mathbb{R}^n$ it is known that the set of all RFDE's in a convenient class of functions $\mathscr{X}$ which have all critical points and all periodic orbits hyperbolic is a residual set in $\mathscr{X}$. These results are described below in detail, since they illustrate the techniques used in the generic theory of RFDE's.

The proof follows the general pattern that was developed for ordinary differential equations. We consider RFDE's on $\mathbb{R}^n$, defined by

$$\dot{x}(t) = f(x_t) \tag{4.1}$$

with $f \in \mathscr{X} = \mathscr{X}^k(I, \mathbb{R}^n)$, $k \geq 2$, and taking $\mathscr{X}^k$ with the C^k-uniform topology. For each compact set $K \subset \mathbb{R}^n$ and each $A > 0$ define the subsets of :

$\mathscr{G}_0(K)$ = {f: all critical points in K are nondegenerate}

$\mathscr{G}_1(K)$ = {f: all critical points in K are hyperbolic}

$\mathscr{G}_{3/2}(K,A)$ = {$f \in \mathscr{G}_1(K)$: all periodic orbits lying in K and with period in $(0,A]$ are nondegenerate}

$\mathscr{G}_2(K,A)$ = {$f \in \mathscr{G}_1(K)$: all periodic orbits lying in K and with period in $(0,A]$ are hyperbolic}.

Theorem 4.1. The set of all $f \in \mathscr{X}$ such that all critical points and all periodic orbits of (4.1) are hyperbolic is residual in $\mathscr{X}$.

Proof: We break the proof in several steps:

1) the sets $\mathscr{G}_0(K)$, $\mathscr{G}_1(K)$, $\mathscr{G}_{3/2}(K,A)$, $\mathscr{G}_2(K,A)$ are open.

This is a consequence of general perturbation results associated with the saddle-point property.

2) $\mathscr{G}_0(K)$ is dense in $\mathscr{X}$.

Any $f \in \mathscr{X}$, by restriction to the constant functions in C^0, gives a C^k function $\bar{f}: \mathbb{R}^n \to \mathbb{R}^n$. It is easily seen that $\varphi \in K$ is a critical point of (4.1) if and only if the origin of $\mathbb{R}^n$ is a regular value of the restriction of $\bar{f}$ to a compact set $\bar{K} \subset \mathbb{R}^n$. By Sard's theorem, the set of singular values of $\bar{f}$ has measure zero, so there are regular values $\varepsilon \in \mathbb{R}^n$ of $\bar{f}$ arbitrarily close to zero. Letting $G: \mathbb{R}^n \to \mathbb{R}^n$ be a C^∞-function with compact support and equal to 1 on $\bar{K}$, and $g_\varepsilon: C^0 \to \mathbb{R}^n$ be defined by $g_\varepsilon(\varphi) = f(\varphi) - \varepsilon G(\varphi(0))$, we get $g_\varepsilon \in \mathscr{X}$ and $\bar{g}_\varepsilon = \bar{f} - \varepsilon G$. Consequently, $g_\varepsilon \in \mathscr{G}_0(K)$ and, therefore, $\mathscr{G}_0(K)$ is dense in $\mathscr{X}$.

3) $\mathscr{G}_1(K)$ is dense in $\mathscr{G}_0(K)$.

Take $f \in \mathscr{G}_0(K)$. Each zero of $\bar{f}$ in $\bar{K}$ is isolated and, since $\bar{K}$ is compact, the zeros of $\bar{f}$ are finitely many. By the Implicit Function Theorem, these zeros persist under small perturbations of f and no other new zeros of $\bar{f}$ appear in some neighborhood of $\bar{K}$. If we can perturb f locally around each critical point, by adding to f a function having support in a small neighborhood of the point, in such a way that the associated critical point of the perturbed equation is hyperbolic, then we can construct perturbations of f which have the same number of critical points on K as f does, but with all of them hyperbolic. This would imply that $\mathcal{G}_1(K)$ is dense in $\mathcal{G}_0(K)$.

To show that such local perturbations exist, let $a \in \overline{K}$ be a zero of $\overline{f}$ and change coordinates so that $a = 0$. Let $H: \mathbb{R}^n \times \mathbb{R}^n \to \mathbb{R}^n$ be C^∞ with arbitrarily small compact support and $H(0,0) = 0$, $DH(0,0) = (0,I)$. For $\varepsilon \in \mathbb{R}$, let $L_\varepsilon: C^0 \to \mathbb{R}^n$ be $L_\varepsilon(\varphi) = -\varphi(0) + [(e^\varepsilon-1)/\varepsilon]f'(0)\varphi$ and define $g_\varepsilon(\varphi) = f(\varphi) - \varepsilon H(\varphi(0), L_\varepsilon(\varphi))$. Then, as $\varepsilon \to 0$, $g_\varepsilon \to f$ in $\mathcal{X}$, and the characteristic function $\Delta_\varepsilon(\lambda)$ of the linearized equation at zero satisfy $\Delta_\varepsilon(\lambda) = \Delta_0(\lambda+\varepsilon)$. For all $\varepsilon \neq 0$ small, Δ_0 has no zeros on $\operatorname{Re}\lambda = \varepsilon$. Thus 0 is an hyperbolic critical point of $\dot{x}(t) = g_\varepsilon(x_t)$.

4) $\mathcal{G}_{3/2}(K,3A/2)$ is dense in $\mathcal{G}_2(K,A)$.

The main idea for proving this statement is to consider, for each $f \in \mathcal{G}_2(K,A)$, perturbations on a conveniently chosen finite-dimensional subspace of $\mathcal{X}$. The elements of this subspace are taken from the set

$$\mathcal{Y} = \{f \in \mathcal{X}: f(\varphi) = F(\varphi(0), \varphi(-r/N), \varphi(-2r/N), \ldots, \varphi(-r))$$
$$\text{for some } N \text{ and some } F \in C^k(\mathbb{R}^{n(N+1)}, \mathbb{R}^n)\}.$$

Fix $f \in \mathcal{G}_2(K,A)$ and $g_1, \ldots, g_J \in \mathcal{Y}$. For $\eta = (\eta_1, \ldots, \eta_J) \in \mathbb{R}^J$, let $f_\eta = f + \sum_{j=1}^{J} \eta_j g_j$, and denote by $x(t;\varphi,\eta)$ the solution of the initial value problem

$$\dot{x}(t) = f_\eta(t), \quad x_0 = \varphi. \tag{4.2}$$

Consider the map $\Psi: (0,\infty) \times C^0 \times \mathbb{R}^J \to C^0$ given by $\Psi(t,\varphi,\eta) = x_t(\varphi,\eta) - \varphi$. Clearly, the zeros of Ψ correspond to initial data of periodic solutions of

(4.2). Let $x^*(t)$ be a nonconstant periodic solution of (4.1) lying in K, having period $t^* \in (0,3A/2]$ and nonconstant initial conditions $\varphi^* = x_0^*$. Then $\Psi(t^*,\varphi^*,0) = 0$. The Implicit Function Theorem cannot be applied to Ψ, since this map may fail to be differentiable at (t,φ,η) if $\varphi \notin C^1$. For this reason, we introduce, for each integer N, the map $\Psi_N(t,\varphi,\eta) = \Psi(Nt,\varphi,\eta)$, which is differentiable at (t,φ,η) provided N is sufficiently large (see Th. 2.2). By application of the Implicit Function Theorem to functions Ψ_N, for conveniently chosen N, we get the following lemma.

Lemma 4.2. *If* $D\Psi(t^*,\varphi^*,0)$ *is surjective, then there is a neighborhood* U *of* $(t^*,\varphi^*,0)$ *in* $(0,\infty) \times C^0 \times \mathbb{R}^J$ *such that* $M = \Psi^{-1}(0) \cap U$ *is a* C^2-*manifold. At each point* $(t,\varphi,\eta) \in M$, $D\Psi(t,\varphi,\eta)$ *is surjective, and the tangent space of* M *is the null space of* $D\Psi(t,\varphi,\eta)$.

Proof of Lemma 4.2: Let $\Lambda = D_\varphi x_{t^*}^*(\varphi^*,0)$, $\Gamma = D_\eta x_{t^*}^*(\varphi^*,0)$ and notice that Λ can be defined by the solution map of the linear variational equation of (4.1) at the periodic solution $x^*(t)$

$$\dot{y}(t) = f'(x_t^*)y_t, \quad y_0 = \xi \tag{4.3}$$

as

$$\Lambda\xi = y_{t^*}. \tag{4.4}$$

A straightforward computation gives

$$D\Psi_N(t^*,\varphi^*,0)(s,\psi,\sigma) = \left(\sum_{i=1}^{N-1} \Lambda^i\right)(\dot{\varphi}^*s+(\Lambda-I)\psi+\Gamma\sigma). \tag{4.5}$$

Since Λ is defined by (4.3)-(4.4), some power of Λ is a compact operator. There are finitely many points of norm one in the spectrum $\sigma(\Lambda)$ of Λ.

Therefore, there exist relatively prime positive integers N_1, N_2 such that $\exp(2\pi ik/N_3) \notin \sigma(\Lambda)$ for all $0 < k < N_3 = N_1 N_2$. Then

$$0 \notin \sigma\left(\sum_{i=0}^{N_j-1} \Lambda^i\right) \quad \text{for} \quad j = 1,2,3$$

and therefore, these operators are isomorphisms. This implies that $D\Psi_{N_j}(t^*,\varphi^*,0)$ $j = 1,2,3$ are surjective and the null spaces of $D\Psi_{N_j}(t^*,\varphi^*,0)$, $j = 1,2,3$ and of $D\Psi(t^*,\varphi^*,0)$ are equal. Let P and Q, $P + Q = I$, be the usual spectral projections onto Λ-invariant subspaces, where $\Lambda - I$ is nilpotent on the finite-dimensional P space, and has an inverse L on the Q space. Noting that $\Lambda\dot{\varphi}^* = \dot{\varphi}^*$ and $P\dot{\varphi}^* = \dot{\varphi}^*$, we see that $(s,\psi,\sigma) \in$ null $(D\Psi)$ is equivalent to the following system

$$\dot{\varphi}^* s + (\Lambda - I)P\psi + P\Gamma\sigma = 0 \tag{4.6}$$

$$Q\Psi = -LQ\Gamma\sigma. \tag{4.7}$$

The only independent parameters in this system are $s, P\psi$ and σ, which are all finite-dimensional. Thus $\text{null}(D\Psi) = \text{null}(D_{N_j}\Psi)$, $j = 1,2,3$ is finite dimensional, and, consequently, has a closed complement. Since $N_j > 2/t^*$, $j = 1,2,3$, it follows that Ψ_{N_j} is C^2 at $(t^*,\varphi^*,0)$. By the Implicit Function Theorem, there is a neighborhood U of $(t^*,\varphi^*,0)$ such that $M_j = \Psi_{N_j}^{-1}(0) \cap U$ is a C^2-manifold, and at each $(t,\varphi,\eta) \in M_j$ the map $D\Psi_{N_j}$ is surjective, and the tangent space of M_j is the null space of $D\Psi_{N_j}(t,\varphi,\eta)$. Any solution of period $N_j t$ $(j = 1,2)$ is also of period $N_3 t$, so $M_j \subset M_3$, $j = 1,2$. Furthermore, the tangent space of M_j at $(t^*,\varphi^*,0)$ is independent of $j = 1,2,3$. So, by restriction to a smaller neighborhood U, if necessary, we get $M_1 = M_2 = M_3 \overset{\text{def}}{=} M_0$. Clearly, $\Psi^{-1}(0) \cap U \subset M_0$. On the other hand, $(t,\varphi,\eta) \in M_0$ is associated with a solution of (4.2) with periods $N_1 t$ and $N_2 t$ and, since N_1, N_2 are relatively prime, this solution has period t and $(t,\varphi,\eta) \in \Psi^{-1}(0) \cap U$. This proves $\Psi^{-1}(0) \cap U = M_0$ is a C^2-manifold. The assertions in Lemma 4.2 about surjectivity and tangent space follow from (4.5), evaluated at (t,φ,η), by noting that $\sum_{i=0}^{N_j-1} (D_\varphi x_y(\varphi,\eta))^i$, $j = 1,2$ are isomorphisms for $(t,\varphi,\eta) \in M_0$, provided U is taken sufficiently small. This finishes the proof of Lemma 4.2.

On the basis of Lemma 4.2, we now need to prove that $D\Psi(t^*,\varphi^*,0)$ is surjective. If $x^*(t)$ is a nondegenerate periodic solution of (4.1) with period t^*, then, the map $(s,\psi) \to \dot{\varphi}^* s + (\Lambda - I)\psi$ from $(0,\infty) \times C^0$ into C^0 is surjective. From (4.5) with $N = 1$, it follows that $D\Psi(t^*,\varphi^*,0)$ is surjective. On the other hand, if $x^*(t)$, with period t^*, is degenerate, then $f \in \mathcal{G}_2(K,A)$ implies $t^* \in [A, 3A/2]$, and, consequently t^* is the least period of $x^*(t)$. The following lemma, whose proof is postponed, guarantees that $D\Psi(t^*,\varphi^*,0)$ is surjective for some choice of $g_1,\dots,g_J \in \mathcal{Y}$.

Lemma 4.3. *If $\bar{t}$ is the least period of a periodic solution of* (4.1) *through $\bar{\varphi}$, then there exist $g_1,\dots,g_J \in \mathcal{Y}$ such that $D\Psi(\bar{t},\bar{\varphi},0)$ is surjective.*

Lemma 4.2 can then be applied at each point (t,φ) in the set $F(0)$, where

$$F(\eta) = \{(t,\varphi);\ 0 < t \le 3A/2,\ \varphi \text{ is not constant, } x(s;\varphi,\eta) \in K \text{ for all } s \in [0,3A/2] \text{ and } \Psi(t,\varphi,\eta) = 0\}.$$

J and $g_1,\dots,g_J \in \mathcal{Y}$ are possibly different for different points (t,φ). To remove this dependence on (t,φ), notice that, given $f \in \mathcal{G}_0(K)$, there exists a neighborhood $\mathcal{N}$ of f such that the periods of nonconstant periodic solutions of $\dot{x}(t) = g(x_t)$, $g \in \mathcal{N}$, lying in K are bounded below by some $\varepsilon > 0$, and, then, observe that $F(\eta)$ is compact. By compactness of $F(0)$, one can find finitely many $g_1,\dots,g_J \in \mathcal{Y}$ such that $D\Psi(t,\varphi,0)$ is surjective for all $(t,\varphi) \in F(0)$. Lemma 4.2 then implies there exists a neighborhood U of $F(0) \times \{0\}$ in $(0,\infty) \times C^0 \times \mathbb{R}^J$ such that the conclusions of that lemma hold for $M = \Psi^{-1}(0) \cap U$.

Now, consider the projection $\pi: M \longrightarrow \mathbb{R}^J$ given by $\pi(t,\varphi,\eta) = \eta$. Since the tangent space of M at (t,φ,η) is equal to the null space of $D\Psi(t,\varphi,\eta)$, and formula (4.5) holds for $N = 1$ at each $(t,\varphi,\eta) \in \Psi^{-1}(0)$ (replacing φ^* by φ and Λ and Γ by the corresponding derivatives computed at (t,φ,η) instead of $(t^*,\varphi^*,0)$) we get

$$\operatorname{null}(D\pi) = \{(s,\psi,0) \in \mathbb{R} \times C^0 \times \mathbb{R}^J : \dot{\varphi}s + (\Lambda - I)\psi = 0\}$$

$$\operatorname{range}(D\pi) = \{\sigma \in \mathbb{R}^J : \dot{\varphi}s + (\Lambda - I)\psi + \Gamma\sigma = 0 \text{ for some } (s,\psi) \in \mathbb{R} \times C^0\}.$$

The reasoning leading to (4.6)-(4.7) is also valid in the present situation and we can compute the dimensions of $\operatorname{null}(D\pi)$ and $\operatorname{range}(D\pi)$ by studying the finite-dimensional equation (4.6) with the use of the surjectivity of the map $(s,P\psi,\sigma) \to \dot{\varphi}_s + (\Lambda - I)P\psi + P\Gamma\sigma$. It is then possible to show that the Fredholm index of $D\pi$, $\dim \operatorname{null}(D\pi) - \operatorname{codim} \operatorname{range}(D\pi)$, is equal to one. Since M and π are of class C^2, we can now apply Smale's version of Sard's theorem to get that the set of regular values of π is residual in

$\mathbb{R}^J$. In particular, there are regular values arbitrarily near zero. On the other hand, the upper semicontinuity of $F(\eta)$, guarantees that $F(\eta) \times \{\eta\} \subset U$ for sufficiently small η. For such η which are regular values of π, we have that any solution of (4.2) of period $\leq 3A/2$, lying in K, must correspond to a point $(t,\varphi,\eta) \subset M$. Since $D\Psi$ is surjective at points of M and η is a regular value of π implies $D\pi(t,\varphi,\eta)\colon \mathrm{null}(D\Psi(t,\varphi,\eta)) \mapsto \mathbb{R}^J$ is also surjective. Thus $(s,\psi) \to D\Psi(t,\varphi,\eta)(s,\psi,0)$ is surjective, implying that the solution of (4.2) with initial condition φ is a nondegenerate periodic solution with period t. This finishes the proof that $\mathcal{G}_{3/2}(K, 3A/2)$ is dense in $\mathcal{G}_2(K,A)$.

5) $\mathcal{G}_2(K,A)$ is dense in $\mathcal{G}_{3/2}(K,A)$.

Fix $f \in \mathcal{G}_{3/2}(K,A)$. Each periodic solution $x^*(t)$ of (4.1) of period $t^* \leq A$ and lying in K is nondegenerate. Therefore, there exists a neighborhood of its orbit which contains no other periodic orbits of period close to t^*, and, under small perturbations of (4.1), the periodic solution and its period depend continuously on the perturbation. If N_1,N_2,N_3 are chosen as in the proof of Lemma 4.2 and since $x^*(t)$ is a periodic nondegenerate solution of (4.1) with any of the periods $N_jt^* > 1$, $j = 1,2,3$, there are unique orbits Γ_j of periods near N_jt^* and changing continuously with the perturbation, for $j = 1,2,3$. The orbits of period near N_jt^*, $j = 1,2$ also have periods near $N_1N_2t^* = N_3t^*$, and, therefore, $\Gamma_1 = \Gamma_2 = \Gamma_3 \overset{\text{def}}{=} \Gamma$. Since N_1,N_2 are relatively prime we have that the period of Γ is near t^* and depends continuously on the perturbation. By compactness, there are only finitely many periodic solutions of (4.1) of periods $\leq A$ and lying in K. To prove that $\mathcal{G}_2(K,A)$ is dense in $\mathcal{G}_{3/2}(K,A)$, it is sufficient to make a small perturbation in a neighborhood of each periodic solution.

Assume $x^*(t)$ is a nondegenerate periodic solution of (4.1) with least period $t^* \leq A$. Let $y^j(t)$, $j = 1,\dots,d$ be solutions of the variational equation of (4.1) which form a basis for the generalized eigenspace corresponding to all characteristic multipliers of $x^*(t)$ having $|\mu| = 1$. Without loss of generality we take $y^1(t) = \dot{x}^*(t)$. Letting $Y(t) = (y^1(t),\dots,y^d(t))$, there exists a $d \times d$ matrix M with all eigenvalues in the unit circle such that $Y(t+t^*) = Y(t)M$. After changing the basis so that M is in Jordan canonical form, it is not difficult to perturb $Y(t)$ and M to a differentiable function $Y^\varepsilon(t)$ and M^ε, for ε small, so that $Y^0 = Y$, $M^0 = M$, and the eigenvalues of M^ε are all off the unit circle except for the eigenvalue 1 which is simple, and $Y^\varepsilon(t+t^*) = Y^\varepsilon(t)M^\varepsilon$. In order to perturb (4.1) as

$$\dot{x}(t) = f(x_t) + g(x_t) \tag{4.8}$$

and have the periodic solution $x^*(t)$ of (4.1) transformed to an hyperbolic periodic solution of (4.8) we can try to choose g so that $x^*(t)$ is still a solution of (4.8) and $Y^\varepsilon(t)$ is a solution of the linear variational equation of (4.8) around $x^*(t)$. It is not difficult to show that this can be accomplished choosing $g \in \mathscr{Y}$. If we denote

$$\delta_N\varphi = (\varphi(0),\varphi(-r/N),\varphi(-2r/N),\dots,\varphi(-r)),$$

then the appropriate functions g are of the form

$$g(\varphi) = G(\delta_N\varphi),$$

with the function $G(x_1,\dots,x_{N+1})$ satisfying

$$\dot{Y}^{\varepsilon}(t) = f'(x_t^*)Y_t^{\varepsilon} + \left[\frac{\partial G}{\partial x_1}, \ldots, \frac{\partial G}{\partial x_{N+1}}\right]_{\delta_N x_t^*} \delta_N Y_t^{\varepsilon}. \tag{4.9}$$

It is, therefore, enough to find a function G satisfying this equation. Assume there exist sequences $t_N \subset [0,T]$, v_N of d vectors with norm smaller or equal to 1, and $\varepsilon_N \to 0$ as $N \to \infty$ such that $Y^{\varepsilon_N}(t_N - k/N)v_N = 0$ for all $0 \leq k \leq N$. Given an arbitrary $\theta \in [-1,0]$, there exists a sequence k_N such that $0 \leq k_N \leq N$ and $k_N/N \to \theta$ as $N \to \infty$. Taking subsequences if necessary, we get $t_N \to \tau$, $v_N \to \omega$, $\varepsilon_N \to 0$, $Y(\tau+\theta)\omega = 0$. Therefore $Y_\tau\omega = 0$ since the columns of Y_τ are linearly independent, we must have $\omega = 0$. Consequently, for N sufficiently large and ε close to zero, the columns of $\delta_N Y_t^{\varepsilon}$ are linearly independent for all $t \in [0,T]$. Since M^{ε} is non-singular and $\delta_N Y_{t+T}^{\varepsilon} = (\delta_N Y_t^{\varepsilon})M^{\varepsilon}$, the columns of $\delta_N Y_t^{\varepsilon}$ are linearly independent for all $t \in \mathbb{R}$, whenever N is large and ε is close to zero. As $\delta_N Y^{\varepsilon}$ is a matrix of dimension $n(N+1) \times d$, the equation (4.9) for the unknown $\left[\frac{\partial G}{\partial x_1}, \ldots, \frac{\partial G}{\partial x_{N+1}}\right]_{\delta_N x_t^*}$ is underdetermined for N large and ε close to zero, and we can get one particular solution by multiplying (4.9) by the Moore-Penrose generalized inverse of $\delta_N Y_t^{\varepsilon}$

$$(\delta_N Y_t^{\varepsilon})^+ = [(\delta_N Y_t^{\varepsilon})^T(\delta_N Y_t^{\varepsilon})]^{-1}(\delta_N Y_t^{\varepsilon})^T$$

where the superscript T denotes transpose. We get

$$\left[\frac{\partial G}{\partial x_1}, \ldots, \frac{\partial G}{\partial x_{N+1}}\right]_{\delta_N x_t^*} = (\dot{Y}^{\varepsilon}(t) - f'(x_t^*)Y_t^{\varepsilon})(\delta_N Y_t^{\varepsilon})^+. \tag{4.10}$$

Since we want equation (4.8) to be a local perturbation of (4.1) around x_t^*, we look for a function G of compact support, small as $\varepsilon \to 0$, vanishing over $\gamma = \{(x^*(t),x^*(t-r/N),\ldots,x^*(-r)) : 0 \leq t \leq t^*\}$ and satisfying (4.10). Choosing a local tubular coordinate system $(u_1,u_2,\ldots,u_{n(N+1)})$ around γ with γ corresponding to $\{u_2 = \ldots = u_{n(N+1)} = 0\}$ and u_1 of period t^*, we must then have

$$G(u_1,0,\ldots,0) = 0, \quad \frac{\partial G}{\partial u_j}(u_1,0,\ldots,0) = \gamma_j(u_1), \quad j = 1,\ldots,n(N+1),$$

where $\gamma_j(u_1)$ are given by the right-hand side of (4.10). Since f is of class C^k, the γ_j are of class C^{k-1}. But, as we want $g \in \mathscr{X} = \mathscr{X}^k$, we need G to be a C^k function. We can achieve this by integral averaging, in order to recover the missing degree of smoothness, as

$$G(u_1,\ldots,u_{n(N+1)}) = \sum_{j=2}^{m} u_j \int_0^\infty \gamma_j(u_1+vu_j)\rho(v)dv$$

where $\rho\colon [0,\infty) \to \mathbb{R}$ is C^∞, has compact support and satisfies $\rho(v) = 1$ for $v \in [0,1]$. After multiplication by a C^∞ bump function of compact support and equal to 1 near γ, we get G such that the perturbation of (4.1) defined by (4.9) is a small local perturbation of (4.1) around x_t^*, with x_t^* being an hyperbolic solution of the perturbed equation (4.8).

By adding such local perturbations around each one of the (finitely many) nonhyperbolic solutions of (4.1) lying in K and having periods in $[0,A]$, we get small perturbations $(f+g) \in \mathscr{G}_2(K,A)$ of f. This finishes the proof that $\mathscr{G}_2(K,A)$ is dense in $\mathscr{G}_{3/2}(K,A)$.

We are now in the situation of being able to use the induction procedure introduced by Peixoto for ordinary differential equations. Since $\mathcal{G}_2(K, 3A/2)$ is dense in $\mathcal{G}_2(K,A)$ (by 4)) and $\mathcal{G}_2(K,A)$ is dense in $\mathcal{G}_{3/2}(K,A)$ (by 5)), it follows by induction that $\mathcal{G}_2(K,A)$ is dense in $\mathcal{G}_2(K,B)$ for all $B < A$. It was mentioned before that, for any $f \in \mathcal{G}_0(K)$, there exists a neighborhood $\mathcal{N}$ of f in $\mathcal{X}$ such that the periods of nonconstant periodic solutions of $\dot{x}(t) = g(x_t)$, $g \in \mathcal{N}$ lying in K are bounded below by some $\varepsilon > 0$. Thus $\mathcal{N} \subset \mathcal{G}_2(K,\varepsilon)$, implying that $\mathcal{G}_2(K,\varepsilon)$ is dense in $\mathcal{X}$ and, thus, also $\mathcal{G}_2(K,A)$ is dense in $\mathcal{X}$ for all A. Since $\mathcal{G}_0(K)$ is dense in $\mathcal{X}$ (by 2)), it follows that $\mathcal{G}_2(K,A)$ is dense in $\mathcal{X}$ for all A. The set

$$\mathcal{G}_2 = \{f \in \mathcal{X}\text{: all critical points and all periodic solutions of (3.1) are hyperbolic)}\}$$

can be expressed as a countable intersection of sets of the form $\mathcal{G}_2(K,A)$ with K compact and $A > 0$. Consequently, $\mathcal{G}_2$ is residual in $\mathcal{X}$, finishing the proof of the theorem.

It remains to prove Lemma 4.3. For this proof, we use the following result:

Lemma 4.4. Let $x^*(t)$ be a periodic solution of (4.1) of least period $t^* > 0$. Then, for sufficiently large N, the map

$$t \longmapsto (x^*(t), x^*(t-r/N), x^*(t-2r/N), \ldots, x^*(t-r))$$

is a one-to-one regular (that is the derivative $\neq 0$ everywhere) mapping of the reals mod t^* into $\mathbb{R}^{n(N+1)}$.

Proof: If the statement is not true, there would exist arbitrarily large N such that either: 1) there are $t_1 \not\equiv t_2 \pmod{t^*}$ with $x^*(t_1-kr/N) = x^*(t_2-kr/N)$ for all $0 \leq k \leq N$, or 2) there is t_3 with $\dot{x}^*(t_3-kr/N) = 0$ for all $0 \leq k \leq N$. Consequently, one could find a sequence of integers $N_m \to \infty$ as $m \to \infty$ with either $t_1(N_m)$ and $t_2(N_m)$ or $t_3(N_m)$ as above, and take convergent subsequences $t_j(N_m) \to \tau_j$ as $m \to \infty$. On the other hand, for any $\theta \in [-r,0]$, there exists a sequence $0 \leq k_m \leq N_m$ such that $-rk_m/N_m \to \theta$ as $m \to \infty$. If 1) holds and $\tau_1 \not\equiv \tau_2 \pmod{t^*}$, then $x^*(\tau_1+\theta) = x^*(\tau_2+\theta)$ for all $\theta \in [-r,0]$, contradicting that t^* is the least period of x^*. If 1) holds and $\tau_1 \equiv \tau_2 \pmod{t^*}$, then

$$(t_1-t_2)^{-1}[x^*(t_1-kr/N)-x^*(t_2-kr/N)] \to \dot{x}(\tau_1+\theta)$$

as $m \to \infty$, and $\dot{x}(\tau_1+\theta) = 0$ since each term in the sequence vanishes. This would imply $\dot{x}^*_{\tau_1} = 0$, a contradiction since x^* is nonconstant. Finally, if 2) holds, then $\dot{x}^*_{\tau_3} = 0$, also a contradiction.

Proof of Lemma 4.3: Let $\delta_N\colon C^0 \to \mathbb{R}^{n(N+1)}$ denote the map

$$\delta_N\varphi = (\varphi(0),\varphi(-r/N),\varphi(-2r/N),\dots,\varphi(-r)).$$

Then $g \in \mathcal{Y}$ is equivalent to $g(\varphi) = G(\delta_N\varphi)$ for some N and some $G \in C^k(\mathbb{R}^{n(N+1)}, \mathbb{R}^n)$ of compact support.

Suppose first that $t^* > r$. Given any $\xi \in C^0$, there is a $z \in C^0([-r,t^*], \mathbb{R}^n)$ such that $z_0 = 0$, $z_{t^*} = \xi$, and there is a $y^* \in C^1([-r,t^*], \mathbb{R}^n)$ with $y^*_0 = 0$ and arbitrarily close to z, uniformly on $[-r,t^*]$. Defining $\gamma^*(t) = \dot{y}^*(t) - f'(x^*_t)y^*_t$ and applying the variation of constants formula, we can get the solution of $\dot{y}(t) = f'(x^*_t)y_t + \gamma(t)$, $y_0 = 0$ arbitrarily close to z, uniformly on $[-r,t^*]$, by choosing γ sufficiently close to γ^* in $L^1(0,t^*)$. By Lemma 4.4, taking N sufficiently large one may define a function G on $\{\delta_N x_t^*;\ t \in \mathbb{R}\}$ by $G(\delta_N x_t^*) = \gamma'(t)$ and then extend G to the whole of $\mathbb{R}^{n(N+1)}$ as a C^k function of compact support to get $\gamma(t) = g(x_t^*)$ for some $g \in \mathcal{Y}$.

Since $t^* > r$, $\Lambda = D_\eta x_{t^*}^*(\varphi^*,0)$ is compact (see theorem (4.1-4)). It follows that the range of $\Lambda - I$ has finite codimension in C^0. Let $\xi_1,\ldots,\xi_J$ be a basis for a linear complement of range$(\Lambda - I)$. By the argument of the preceding paragraph, one can get the value of the solution $y^i{}_{t^*}$ of $\dot y(t) = f'(x_t^*)y_t + g_i(x_t^*)$, $y_0 = 0$, arbitrarily close to ξ_i, by choosing $g_1,\ldots,g_J$ appropriately in $\mathcal{Y}$. The approximation of the ξ_i by the $y^i_{t^*}$ can be made so close that $y^1_{t^*},\ldots,y^J_{t^*}$ form a basis for a linear complement of range$(\Lambda - I)$. Using the notation on the proof of theorem (4.1-4) with this choice of g_j in the definition of $f_\eta(\varphi) = f(\varphi) + \sum_{j=1}^{J} \eta_j g_j(\varphi)$, it is clear that $\Gamma = D_\eta x_{t^*}^*(\varphi^*,0)$ can be defined in terms of the solution of

$$\dot y(t) = f'(x_t^*)y_t + g(x_t^*), \quad y_0 = 0 \tag{4.11}$$

as

$$\Gamma\sigma = y_{t^*} \quad \text{with} \quad g = \sum_{j=1}^{J} \sigma_j g_j. \tag{4.12}$$

Thus, for $\sigma^i = (\delta_1^i,\ldots,\delta_J^i)$ with δ_j^i the Kronecker delta, we have $\Gamma\sigma^i = y^i_{t^*}$, $i = 1,\ldots,J$. It follows that the map $(\psi,\sigma) \longmapsto (\Lambda - I)\psi + \Gamma\sigma$ is surjective, and therefore, by equation (4.5), $D\Psi(t^*,\varphi^*,0)$ is also surjective.

Now suppose $0 < t^* \le r$ and consider the problem

$$(\Lambda - I)\psi + \Gamma\sigma = \xi \tag{4.13}$$

for $\xi \in C^0$ given. Since $\Gamma\sigma$ satisfies (4.11-4.12), and Λ satisfies (4.3-4.4), this equation is equivalent to the system

$$\psi(t^*+\theta) - \psi(\theta) = \xi(\theta), \quad -r \leq \theta \leq -t^* \tag{4.14}$$

$$[(\Lambda-I)\psi + \Gamma\sigma](\theta) = \xi(\theta), \quad -t^* \leq \theta \leq 0. \tag{4.15}$$

The general solution of (4.14) is $\psi = \psi_0 + \psi_1$ where ψ_0 is a particular solution of the equation and $\psi_1 \in C^0$ is any function of period t^*. Fixing ψ_0, (4.15) becomes

$$[(\Lambda-I)\psi_1+\Gamma\sigma](\theta) = \xi_1(\theta), \quad -t^* \leq \theta \leq 0 \tag{4.16}$$

where $\xi_1 = \xi - (\Lambda-I)\psi_0$. Let C_p be the space of the t^*-periodic continuous functions of $[-t^*,0]$ into $\mathbb{R}^n$, and let $L: C_p \to C([-t^*,0], \mathbb{R}^n)$ assign to each element $\psi_1 \in C_p$ the constant function with value $\psi_1(0)$. Then

$$V = \{(\Lambda-I)\psi_1\big|_{[-t^*,0]} : \psi_1 \in C^0 \text{ is } t^*\text{-periodic}\}$$

is equal to

$$\{(\Lambda-L)\psi_1\big|_{[-t^*,0]} + (L-I)\psi_1\big|_{[-t^*,0]} : \psi_1 \in C_p\}.$$

Since $(\Lambda-L)|_{[-t^*,0]}$ is compact (for the same reason that Λ is when $t^* > 1$) and $(\Lambda-I)|_{[-t^*,0]}$ is an isomorphism identifying C_p and

$$C_0 = \{\varphi \in C([-t^*,0], \mathbb{R}^n); \varphi(-t^*) = 0\},$$

it follows that V has finite codimension in C_0. Noting that $\xi_1(-t^*) = 0$, we can proceed as for $t^* > 1$ to get $g_1,\ldots,g_J \in \mathcal{Y}$ such that the map $(\psi,\sigma) \longmapsto (\Lambda-I)\psi + \Gamma\sigma$ is surjective, implying that $D\Psi(t^*,\varphi^*,0)$ is also surjective and finishing the proof of the lemma.

It is interesting to restrict the class of functions $\mathscr{X}$; for example, to consider only differential difference equations of the form

$$\dot{x}(t) = F(x(t), x(t-1)). \qquad (4.16)$$

To obtain a generic theorem about this restricted class of equations is more difficult since there is less freedom to construct perturbations. For example, the functions $g \in \mathscr{Y}$ used in the proof of Theorem 4.1 cannot be used in the present case. Nevertheless, Theorem 4.1 still holds for these equations. The proof of this fact follows the same general scheme as the proof of Theorem 4.1, but the proofs of denseness of $\mathscr{G}_{3/2}(K, 3A/2)$ in $\mathscr{G}_2(K,A)$ and of $\mathscr{G}_2(K,A)$ in $\mathscr{G}_{3/2}(K,A)$ are very different. The role played by Lemma 4.4 in the construction of the perturbations of (4.1) used in the proof of the denseness of $\mathscr{G}_{3/2}(K, 3A/2)$ in $\mathscr{G}_2(K,A)$ is now played by the following lemma after approximating F by an analytic function.

Lemma 4.5. *If $x(t)$ is a periodic solution of Equation (4.16) of least period $t^* > 0$, and F is analytic, then the map*

$$y(t) = (x(t), x(t-1))$$

is one-to-one and regular except at a finite number of t values in the reals mod t^*.

Proof: It can be proved that x is analytic. Thus, any self-intersection of $y(t)$ is either isolated or forms an analytic arc. In the latter case, there exists an analytic function σ defined in an interval I with $\dot{\sigma}(t) \neq 0$ and $\sigma(t) \neq t$ such that $y(t) = y(\sigma(t))$, $x(t) = x(\sigma(t))$. Thus,

$$\dot{x}(t) = F(y(t)) = F(y(\sigma(t))) = \dot{x}(\sigma(t)).$$

By differentiation, we get $\dot{x}(t) = x(\sigma(t))\dot{\sigma}(t)$, implying that $\dot{\sigma}(t) \equiv I$. Hence for some τ, $x(t) = x(t+\tau)$ for $t \in I$ and thus, by analyticity, for all t. Therefore A is a multiple of t^* and the lemma is proved.

One may consider an even more restrictive class of equations of the form

$$\dot{x}(t) = F(x(t-1)).$$

The analogue of Theorem 4.1 for this class is still an open question, since the generic properties of periodic solutions of these equations have not been established.

5. Invariant Sets, Limit Sets and the Attractor

A function $y(t)$ is said to be a global solution of an RFDE(F) on M, if it is defined for $t \in (-\infty,+\infty)$ and, for every $\sigma \in (-\infty,+\infty)$, $x_t(\sigma,y_\sigma,F) = y_t$, $t \geq \sigma$. The constant and the periodic solutions are particular cases of global solutions. The solutions with initial data in unstable manifolds of equilibrium points or periodic orbits are often global solutions, for example, when M is compact. An invariant set of an RFDE(F) on a manifold M, is a subset S of $C^0 = C^0(I,M)$ such that for every $\varphi \in S$ there exists a global solution x of the RFDE, satisfying $x_0 = \varphi$ and $x_t \in S$ for all $t \in \mathbb{R}$. The ω-limit set $\omega(\varphi)$ of an orbit $\gamma^+(\varphi) = \{\Phi_t\varphi,\ t \geq 0\}$ through φ is the set

$$\omega(\varphi) = \bigcap_{\tau\geq 0} C\ell \bigcup_{t\geq\tau} \Phi_t\varphi. \tag{5.1}$$

This is equivalent to saying that $\psi \in \omega(\varphi)$ if and only if there is a sequence $t_n \to \infty$ as $n \to \infty$ such that $\Phi_{t_n}\varphi \to \psi$ as $n \to \infty$. For any set $S \subset C^0$, one can define

$$\omega(S) = \bigcap_{\tau\geq 0} C\ell \bigcup_{\substack{t\geq\tau \\ \varphi\in S}} \Phi_t\varphi.$$

In a similar way, if $x(t,\varphi)$ is a solution of the RFDE(F) for $t \in (-\infty,0]$, $x_0(\cdot,\varphi) = \varphi$, one can define the α-limit set of the negative orbit $\{x_t(\cdot,\varphi),\ -\infty < t \leq 0\}$. Since the map Φ_t may not be one-to-one, there may be other negative orbits through φ and, thus, other α-limit points. To take into account this possibility, we define the α-limit set of φ in the following way. For any $\varphi \in C^0$ and any $t \geq 0$, let

$$H(t,\varphi) = \{\psi \in C^0: \text{there is a solution } x(t,\varphi) \text{ of the RFDE(F) on } (-\infty,0],\ x_0(\cdot,\varphi) = \varphi,\ x_{-t}(\cdot,\varphi) = \psi\}$$

and define the α-limit set $\alpha(\varphi)$ of φ as

$$\alpha(\varphi) = \bigcap_{\tau \geq 0} \mathrm{C}\ell \bigcup_{t \geq \tau} H(t,\varphi) \tag{5.2}$$

Lemma 5.1. *Let* $F \in \mathscr{X}^k$, $k \geq 1$, *be a RFDE on a connected manifold* M. *Then the* ω-*limit set* $\omega(\varphi)$ *of any bounded orbit* $\gamma^+(\varphi)$, $\varphi \in M$ *is non-empty, compact, connected and invariant. The same conclusion is valid for* $\omega(S)$ *for any connected set* $S \subset C^0$ *for which* $\gamma^+(S)$ *is bounded.*

If $\bigcup_{t \geq 0} H(t,\varphi)$ *is non-empty and bounded, then the* α-*limit set* $\alpha(\varphi)$ *is nonempty, compact and invariant. If, in addition,* $H(t,\varphi)$ *is connected, then* $\alpha(\varphi)$ *is connected.*

Remark 1. It seems plausible that $H(t,\varphi)$ is always connected, but it is not known if this is the case.

Remark 2. If M is a compact manifold, then $\gamma^+(\varphi)$, $\bigcup_{t \geq 0} H(t,\varphi)$ are bounded sets and, thus, the ω-limit set is nonempty, compact, connected and invariant. The α-limit set is compact and invariant, being connected if $H(t,\varphi)$ is connected and nonempty if $\bigcup_{t \geq 0} H(t,\varphi)$ is nonempty.

Remark 3. If Φ_t is one-to-one, then $H(t,\varphi)$ is empty or a singleton for each $t \geq 0$ and, thus, the boundedness of the negative orbit of φ implies $\alpha(\varphi)$ is a nonempty, compact, connected invariant set.

Proof of Lemma 5.1: The proof given here follows the proof of the analogous statement for dynamical systems defined on a Banach space. However, in order to emphasize the ideas behind the result, a direct proof is given.

Let $\gamma^+(\varphi) = \{\Phi_t\varphi,\ t \geq 0\}$ be bounded. Since $F \in \mathscr{X}^1$, Ascoli's Theorem can be used to show that $\gamma^+(\varphi)$ is precompact. It follows now directly from the definition of $\omega(\varphi)$ in (5.1) that it is nonempty and compact.

Assume now that $\mathrm{dist}(\Phi_t\varphi,\omega(\varphi)) \not\to 0$ as $t \to \infty$, where dist stands for the admissible metric in $C^0(I,M)$. Then there exist $\varepsilon > 0$ and a sequence $t_k \to \infty$ as $k \to \infty$ such that $\mathrm{dist}(\Phi_{t_k}\varphi\ ,\ \omega(\varphi)) > \varepsilon$ for $k = 1,2,\ldots$. Since the sequence $\{\Phi_{t_k}\varphi\}$ is in a compact set, it has a convergent subsequence. The limit necessarily belongs to $\omega(\varphi)$, contradicting $\mathrm{dist}(\Phi_{t_k}\varphi,\omega(\varphi)) > \varepsilon$. Thus, $\mathrm{dist}(\Phi_t\varphi,\omega(\varphi)) \to 0$ as $t \to \infty$. If $\omega(\varphi)$ were not connected, it would be a union of two disjoint compact sets which would be a distance $\sigma > 0$ apart. This contradicts $\mathrm{dist}(\Phi_t\varphi,\omega(\varphi)) \to 0$ as $t \to \infty$, and so $\omega(\varphi)$ is connected.

Suppose $\psi \in \omega(\varphi)$. There exists a sequence $t_k \to \infty$ as $k \to \infty$ such that $\Phi_{t_k}\varphi \to \psi$. For any integer $N \geq 0$, there exists an integer $k_0(N)$ such that $\Phi_{t_k+t}\varphi$ is defined for $-N \leq t < +\infty$ if $k \geq k_0(N)$. Since $\gamma^+(\varphi)$ is precompact, one can find a subsequence $\{t_{k,N}\}$ of $\{t_k\}$ and a continuous function $y\colon [-N,N] \to \omega(\varphi)$ such that $\Phi_{t_{k,N}+t}\varphi \to y(t)$ as $k \to \infty$, uniformly for $t \in [-N,N]$. By the diagonalization procedure, there exists a subsequence, denoted also by $\{t_k\}$, and a continuous function $y\colon (-\infty,\infty) \to \omega(\varphi)$, such that $\Phi_{t_k+t}\varphi \to y(t)$ as $k \to \infty$, uniformly on compact sets of $(-\infty,+\infty)$. Clearly, $y(t)$, $t \geq \sigma$ is the solution of the RFDE(F) with initial condition y_σ at $t = \sigma$, i.e., $y(t) = x(t;\sigma,y_\sigma,F)$, $t \geq \sigma$. Thus, y is a global solution of the RFDE(F). On the other hand $y(0) = \psi$. Consequently, $\omega(\varphi)$ is invariant.

The assertions for $\omega(S)$, $S \subset M$, which are contained in the statement can now be easily proved, and the assertions relative to $\alpha(\varphi)$, $\varphi \in M$ are proved in an analogous way.

Given an RFDE(F) on M, we denote by <u>A(F)</u> the set of all initial data of global bounded solutions of F. The set A(F) is clearly an invariant set of F. If $F \in \mathscr{X}^1$ and $\gamma^+(\varphi)$ (or $\cup_{t\geq 0} H(t,\varphi)$) is bounded, then Lemma 5.1 implies that $\omega(\varphi)$ (or $\alpha(\varphi)$) is contained in A(F). Consequently, if $F \in \mathscr{X}^1$, the set A(F) contains all the information about the limiting behaviour of the bounded orbits of the RFDE(F). It is important to know when the set A(F) is compact for, in this case, it is the <u>maximal compact invariant set</u> of F. A very simple condition of stability at $t = +\infty$ implies A(F) is compact; namely, point dissipativeness. This condition can be expressed in terms of attractivity properties of sets. For any set $S \subset C^0$ and $\varepsilon > 0$, let $\mathscr{B}(S,\varepsilon) = \{\varphi \in C^0 : \text{dist}(S,\varphi) < \varepsilon\}$, where <u>dist</u> corresponds to the distance measured in the admissible metric of the manifold $C^0 = C^0(I,M)$. We say that a set $S \subset C^0$ <u>attracts a set</u> $U \subset C^0$ (under the RFDE(F)) if, for any $\varepsilon > 0$, there is a $t_0 = t_0(U,\varepsilon)$ such that $\Phi_t U \subset \mathscr{B}(S,\varepsilon)$ for $t \geq t_0$; S is said to be a <u>global attractor</u> if it attracts all points of C^0, i.e., all singletons $\{\varphi\}$, $\varphi \in C^0$. An RFDE(F) on M is said to be <u>point dissipative</u> if there exists a bounded set B which is a global attractor. If F is point dissipative, then besides A(F) being the maximal compact invariant set of F, it also has strong stability properties. In order to discuss these properties, we introduce some more terminology. A set $S \subset C^0$ is said to be stable if, for any $\varepsilon > 0$, there is a $\delta > 0$ such that $\Phi_t \mathscr{B}(S,\delta) \subset \mathscr{B}(S,\varepsilon)$ for $t \geq 0$; S is said to be uniformly asymptotically stable if it is stable and attracts $\mathscr{B}(S,\varepsilon_0)$ for some $\varepsilon_0 > 0$.

In the following, we say Φ_t is a bounded map uniformly on compact subsets of $[0,\infty)$ if, for any bounded set $B \subset C^0$ and any compact set $K \subset [0,\infty)$, the set $\cup_{t \in K} \Phi_t B$ is bounded.

Sometimes we deal with discrete dynamical systems, that is, iterates of a map. In this case, the above concepts are defined in the same way.

Lemma 5.2. If $F \in \mathscr{X}^1$ is a point dissipative RFDE on M and the corresponding solution map, Φ_t, is a bounded map uniformly on compact subsets of $[0,\infty)$, then there is a compact set $K \subset C^0$ which attracts all compact sets of C^0. The set $\mathscr{J} = \bigcap_{n \geq 0} \Phi_{nr} K$ is the same for all compact sets K which attract compact sets of C^0, it is nonempty, compact, connected, invariant and is the maximal compact invariant set.

Proof: Assume the hypotheses in the statement hold and fix $\varepsilon > 0$. Since F is point dissipative, there exists a bounded set B such that, for each $\varphi \in C^0$, there is a $t_0 = t_0(\varphi)$ such that $\Phi_t \varphi \subset \mathscr{B}(B,\varepsilon)$ for $t \geq t_0(\varphi)$. By continuity, for each $\varphi \in C^0$ there is a neighborhood O_φ of φ in M such that $\Phi_t O_\varphi \subset \mathscr{B}(B,\varepsilon)$ for $t_0(\varphi) \leq t \leq t_0(\varphi)+r$. Since, by Theorem 2.3, Φ_r is a compact map, it follows that $B^* = \Phi_r \mathscr{B}(B,\varepsilon)$ is a precompact set and $\Phi_{t+r} O_\varphi \subset B^*$ for $t_0(\varphi) \leq t \leq t_0(\varphi)+r$. If H is an arbitrary compact set of C^0, one can form a finite covering $\{O_{\varphi_i}(H)\}$ with $\varphi_i \in H$ and define $N(H)$ to be the smallest integer greater or equal than $\max_i \{1 + t_0(\varphi_i)/r\}$. Let $H_0 = \cup_i O_{\varphi_i}(H)$ and let $K = \bigcup_{i=0}^{N(B^*)} \Phi_{ir} B^*$. The set K is compact. It is then easy to show that $\Phi_{nr} B^* \subset K$ for $n \geq N(B^*)$ and $\Phi_t H \subset \Phi_t H_0 \subset K$

for $t \geq (N(B^*) + N(H))r$. Consequently, the compact set K attracts all compact sets of C^0.

Applying the above argument to the compact set K itself, we get $\Phi_t K \subset K$ for $t \geq (N(K) + N(B^*))r$. Therefore $\omega(K) \subset K$. Let $\mathcal{J} = \cap_{n \geq 0} \Phi_{nr} K$. Clearly, $\mathcal{J}$ is compact and $\mathcal{J} \subset \omega(K)$. On the other hand, if $\psi \in \omega(K)$ there are sequences $t_j \to \infty$ as $j \to \infty$ and $\varphi_j \in K$ such that $\Phi_{t_j} \varphi_j \to \psi$ as $j \to \infty$. Since $\{\Phi_t K, t \geq (N(K) + N(B^*))r\}$ is precompact, for any integer i one can find a subsequence of $\{\Phi_{t_j - ir} \varphi_j\}$ which converges to some $\psi_i \in \omega(K) \subset K$, and then $\Phi_{ir} \psi_i = \psi$ for all integer i, implying that $\psi \in \mathcal{J}$. This proves $\omega(K) \subset \mathcal{J}$ and, consequently, $\omega(K) = \mathcal{J}$. From Lemma 5.1, $\mathcal{J}$ is nonempty, compact, connected and invariant.

To prove that $\mathcal{J}$ is the maximal compact invariant set, suppose H is any compact invariant set. Since K attracts H and H is invariant, it follows that $H \subset \Phi_{nr} K$ and, therefore, $H \subset \mathcal{J}$.

It remains to prove that $\mathcal{J}$ is independent of the choice of the compact set K which attracts all compact sets of C^0. For this, denote $\mathcal{J} = \mathcal{J}(K)$ and $\mathcal{J}(K_1) = \cap_{n \geq 0} \Phi_{nr} K_1$ where K_1 is a compact set which attracts all compact sets of C^0. Both $\mathcal{J}(K)$ and $\mathcal{J}(K_1)$ are invariant and compact, and they are attracted by both K and K_1. Therefore $\mathcal{J}(K) \subset K_1$, $\mathcal{J}(K_1) \subset K$ and $\mathcal{J}(K) \subset \Phi_{nr} K_1$, $\mathcal{J}(K_1) \subset \Phi_{nr} K$ for all $n \geq 0$. Consequently, $\mathcal{J}(K) = \mathcal{J}(K_1)$.

Remark 4. The above result can also be obtained under weaker hypotheses on Φ_t; namely, it is not necessary to assume that Φ_t takes bounded sets to bounded sets.

Theorem 5.3. If $F \in \mathscr{X}^1$ is a point dissipative RFDE on a connected manifold M and the corresponding solution map, Φ_t, is uniformly bounded on compact subsets of $[0,\infty)$, then A(F) is the maximal compact invariant set of F, it is connected, uniformly asymptotically stable, attracts all bounded sets of C^0 and

$$A(F) = \bigcap_{n \geq 0} \Phi_{nr} K$$

where K is any compact set which attracts all compact sets of C^0.

Proof: By Lemma 5.2, there is a compact set K which attracts all compact sets of C^0, and the set $\mathscr{J} = \bigcap_{n \geq 0} \Phi_{nr} K$ is the maximal compact invariant set of F. Obviously, $\mathscr{J} \subset A(F)$. If

$$A_\beta(F) = \{\varphi \in C^0\text{: there is a global solution } x \text{ of } F \text{ such that } x_0 = \varphi \text{ and } |x_t| \leq \beta \text{ for } t \in (-\infty,\infty)\},$$

then $A(F) = \cup_{\beta>0} A_\beta(F)$. Each one of the sets $A_\beta(F)$ is invariant and, since $F \in \mathscr{X}^1$, the Ascoli's theorem implies $A_\beta(F)$ is compact. By the maximality of $\mathscr{J}$, we have $A_\beta(F) \subset \mathscr{J}$. Consequently, $A(F) = \mathscr{J}$, and all the properties established for $\mathscr{J}$ in Lemma 5.2 also hold for A(F). Given a bounded set $B \subset C^0$, we have $\Phi_r B$ precompact and since A(F) attracts all compact sets of C^0, it also attracts B. It remains to prove that A(F) is stable, since then, as it attracts all bounded sets of C^0, A(F) will be uniformly asymptotically stable.

Let us suppose that $A(F) = \mathscr{J}$ is not stable. Then, for some $\epsilon > 0$ arbitrarily small, there are sequences $\{t_j\}$, $\{\delta_j\} \subset \mathbb{R}^+$, $\{\varphi_j\} \subset C^0$ such that $t_j \to \infty$, $\delta_j > 0$, $\varphi_j \to \mathcal{J}$ as $j \to \infty$, $\Phi_t\varphi_j \in \mathcal{B}(\mathcal{J},\varepsilon)$ for $0 \le t < t_j$ and $\Phi_t\varphi_j \notin \mathcal{B}(\mathcal{J},\varepsilon)$ for $t_j < t < t_j+\delta_j$. Since $\mathcal{J}$ is compact we may assume, without loss of generality, that $\varphi_j \to \varphi \in \mathcal{J}$ as $j \to \infty$. The set $H = \{\varphi,\varphi_j : j \ge 1\}$ is compact and, since $\mathcal{J}$ attracts all compact sets, one has

$$\bigcup_{t \ge T} \Phi_t H \subset \mathcal{B}(\mathcal{J},\varepsilon) \quad \text{for some} \quad T > 0,$$

and, therefore, Lemma 5.1 implies $\omega(H)$ is nonempty, compact and invariant. Since $\mathcal{J}$ is maximal relative to these properties, we have $\omega(H) \subset \mathcal{J}$. As the set $\cup_{t \ge T} \Phi_t H$ is bounded, it follows that $\cup_{t \ge T+r} \Phi_t H$ is precompact. Consequently, the sequence $\{\Phi_{t_j+\delta_j/2}\varphi_j\}_j$ has a subsequence converging to some $z \in \omega(H) \subset \mathcal{J}$. But, by the choice of the t_j, φ_j and δ_j, $z \notin \mathcal{B}(\mathcal{J},\varepsilon)$ which is a contradiction. Therefore, the set $A(F) = \mathcal{J}$ is stable.

Corollary 5.4. If $F \in \mathcal{X}^1$ is an RFDE on a connected compact manifold M, then $A(F)$ is the maximal compact invariant set of F, it is connected, uniformly asymptotically stable, attracts all bounded sets of C^0 and

$$A(F) = \bigcap_{n \ge 0} \Phi_{nr}(C^0)$$

Proof: Noting that $K = C\ell\ \Phi_r(C^0)$ is a compact set (attracting C^0), the corollary is an obvious consequence of Theorem 5.3.

Due to the above properties of the set $A(F)$, it is natural to call it the <u>attractor set</u> of F. Most of the following sections are dedicated to studying properties of this set.

The set A(F) has certain continuity properties in relation to the dependence on F. If M is compact, we have the following theorem, and if M is not compact, some additional hypotheses are needed to obtain a similar result.

<u>Theorem 5.5.</u> *If* $F \in \mathscr{X}^1$ *is an RFDE on a compact manifold* M, *then the attractor set* A(F) *is upper semicontinuous in* F; *that is, for any neighborhood* U *of* A(F) *in* M, *there is a neighborhood* V *of* F *in* $\mathscr{X}^1$ *such that* $A(G) \subset U$ *if* $G \in V$.

<u>Proof</u>: By Corollary 5.4, the attractor A(F) is uniformly asymptotically stable. General results in the theory of stability, based on the construction of "Liapunov functions" guarantee that, for any neighborhood U of A(F) in C^0, there is a neighborhood V of F in $\mathscr{X}^1$ and a $T > 0$ such that the solution map associated with the RFDE $G \in V$, Φ_t^G, satisfies $\Phi_t^G C^0 \subset U$, for all $G \in V$, $t \geq T$. Since, from Lemma 5.4, $A(G) = \bigcap_{n\geq 0} \Phi_{nr}^G(C^0)$, it follows that $A(G) \subset U$.

The preceding argument requires the use of converse theorems on asymptotic stability, establishing the existence of "Liapunov functions". An alternative proof can be given as follows. By Corollary 5.4, the set A(F) is compact and attracts C^0. Let U denote an arbitrarily small neighborhood of A(F), say consisting of all points at a distance from A(F) smaller than a certain $\varepsilon > 0$. Based on Gronwall's inequality one can show that $\Phi_t^F(\varphi)$ and $\Phi_t^G(\varphi)$ can be made as close as desired, uniformly in $\varphi \in C^0$ and $G \in V \subset \mathscr{X}^1$, by choosing V to be a sufficiently small neighborhood of F in $\mathscr{X}^1$. Since $A(F)$ attracts C^0, denoting by W the neighborhood of $A(F)$ consisting of points at a distance from $A(F)$ smaller than $\varepsilon/2$, it follows that there is an integer $N > 0$ such that $\Phi^F_{nr}(C^0) \subset W$ for $n \geq N$. By choosing V sufficiently small we have $\Phi^G_{Nr}(C^0) \subset U$ for all $G \in V$. Since, by Corollary 5.4, $A(G) = \bigcap_{n\geq 0} \Phi^G_{nr}(C^0)$, it follows that $A(G) \subset U$.

Remark 5. The second proof given for the preceding theorem does not generalize for manifolds M which are not compact. However, the first proof can be used, together with some additional hypothesis, to establish a similar result for M not compact.

It has been useful in the generic theory of dynamical systems to consider sets of recurrent motions, in particular, sets of nonwandering points. For an RFDE(F) on a manifold M, an element $\psi \in A(F)$ is called a nonwandering point of F if, for any neighborhood U of ψ in $A(F)$ and any $T > 0$, there exists $t = t(U,T) > T$ and $\tilde{\psi} \in U$ such that $\Phi_t\tilde{\psi} \in U$. The set of all nonwandering points of F is called the nonwandering set of F and is denoted by $\Omega(F)$.

Proposition 5.6. *If $F \in \mathscr{X}^1$ is a point dissipative RFDE on a manifold M, then $\Omega(F)$ is closed and, moreover, if Φ_r is one-to-one on $A(F)$, then $\Omega(F)$ is invariant.*

Proof: The proof follows ideas similar to the ones used in the proof of Lemma 5.1.

Corollary 5.7. *If* $F \in \mathscr{X}^1$ *is an RFDE on a compact manifold* M, *then* $\Omega(F)$ *is closed and, moreover, if* Φ_r *is one-to-one on* $A(F)$, *then* $\Omega(F)$ *is invariant.*

Most of the results in this section are valid in a more abstract setting. We state the results without proof, for maps, and the extension to flows is easy to accomplish.

Throughout the discussion X is a complete metric space and $T: X \to X$ is continuous. The map T is said to be *asymptotically smooth* if for some bounded set $B \subseteq X$, there is a compact set $J \subseteq X$ such that, for any $\varepsilon > 0$, there is an integer $n_0(\varepsilon,B) > 0$ such that, if $T^n x \in B$ for $n \geq 0$, then $T^n x \in (J,\varepsilon)$ for $n \geq n_0(\varepsilon,B)$ where (J,ε) is the ε-neighborhood of J.

Theorem 5.8. *If* $T: X \to X$ *is continuous and there is a compact set* K *which attracts compact sets of* X *and* $J = \cap_n T^n K$, *then*

(i) J *is independent of* K;

(ii) J *is maximal, compact, invariant;*

(iii) J *is stable and attracts compact sets of* X.

If, in addition, T *is asymptotically smooth, then*

(iv) *for any compact set* $H \subseteq X$, *there is a neighborhood* H_1 *of* H *such that* $\cup_{n\geq 0} T^n H_1$ *is bounded and* J *attracts* H_1. *In particular,* J *is uniformly asymptotically stable.*

The following result is useful in the verification of the hypotheses of Theorem 5.8 and, in addition, gives more information about the strong attractivity properties of the set J.

Theorem 5.9. *If* T *is asymptotically smooth and* T *is compact dissipative, then there exists a compact invariant set which attracts compact sets and the conclusions of Theorem 5.8 hold. In addition, if* $\bigcup_{n\geq 0} T^n B$ *is bounded for every bounded set* B *in* X, *then* J *attracts bounded sets of* X.

We now define a more specific class of mappings which are asymptotically smooth.

A *measure of noncompactness* β on a metric space X is a function β from the bounded sets of X to the nonnegative real numbers satisfying

(i) $\beta(A) = 0$ for $A \subseteq X$ if and only if A is precompact,

(ii) $\beta(A \cup B) = \max[\beta(A), \beta(B)]$.

A classical measure of noncompactness is the *Kuratowskii measure of noncompactness* α defined by

$$\alpha(A) = \inf\{d: A \text{ has a finite cover of diameter } < d\}.$$

A continuous map $T: X \to X$ is a β-*contraction of order* $k < 1$ with respect to the measure of noncompactness β if $\beta(TA) \leq k\beta(A)$ for all bounded sets $A \subseteq X$.

Theorem 5.10. β-*contractions are asymptotically smooth.*

From Theorem 5.10 and Theorem 5.9, it follows that T being a β-contraction which is compact dissipative with positive orbits of bounded sets bounded implies there exists a maximal compact invariant set J which attracts bounded sets of X.

It is also very important to know how the set J depends on the map T; that is, a generalization of Theorem 5.5. To state the result, we need another definition.

Suppose $T: \Lambda \times X \to X$ is continuous. Λ and X are complete metric spaces. Also suppose $T(\lambda, \cdot): X \to X$ has a maximal compact invariant set

$J(\lambda)$ for each $\lambda \in \Lambda$. We say $T: \Lambda \times X \to X$ is _collectively β-contracting_ if, for all bounded sets $B, \beta(B) > 0$, one has $\beta(\cup_{\lambda\in\Lambda} T(\lambda,B)) < \beta(B)$.

Theorem 5.11. _Let_ X,Λ _be complete metric spaces_, $T: \Lambda \times X \to X$ _continuous and suppose there is a bounded set_ B _independent of_ $\lambda \in \Lambda$ _such that_ B _is compact dissipative under_ $T(\lambda,\cdot)$ _for every_ $\lambda \in \Lambda$. _If_ T _is collectively_ β-_contracting, then the maximal compact invariant set_ $J(\lambda)$ _of_ $T(\lambda,\cdot)$ _is upper semicontinuous in_ λ.

6. The Dimension of the Attractor

The purpose of this section is to present results on the "size" of the attractor $A(F)$, $F \in \mathscr{X}^k$, $k \geq 1$. This will be given in terms of limit capacity and Hausdorff dimension. The principal results are applicable not only to RFDE's but to the abstract dynamical systems considered in Section 1.

Let K be a topological space. We say that K is <u>finite dimensional</u> if there exists an integer n such that, for every open covering $\mathfrak{A}$ of K, there exists another open covering $\mathfrak{A}'$ refining $\mathfrak{A}$ such that every point of K belongs to at most $n+1$ sets of $\mathfrak{A}'$. In this case, the <u>dimension</u> of K, $\dim K$, is defined as the minimum n satisfying this property. Then $\dim \mathbb{R}^n = n$ and, if K is a compact finite dimensional space, it is homeomorphic to a subset of $\mathbb{R}^n$ with $n = 2 \dim K + 1$. If K is a metric space, its <u>Hausdorff dimension</u> is defined as follows: for any $\alpha > 0$, $\varepsilon > 0$, let

$$\mu_\varepsilon^\alpha(K) = \inf \sum_i \varepsilon_i^\alpha$$

where the inf is taken over all coverings $B_{\varepsilon_i}(x_i)$, $i = 1,2,\ldots$ of K with $\varepsilon_i < \varepsilon$ for all i, where $B_{\varepsilon_i}(x_i) = \{x: d(x,x_i) < \varepsilon_i\}$. Let $\mu^\alpha(K) = \lim_{\varepsilon\to 0} \mu_\varepsilon^\alpha(K)$. The function μ^α is called the <u>Hausdorff measure of dimension</u> α. For $\alpha = n$ and K a subset of $\mathbb{R}^n$ with $|x| = \sup|x_j|$, μ^n is the Lebesgue exterior measure. It is not difficult to show that, if $\mu^\alpha(K) < \infty$ for some α, then $\mu^{\alpha_1}(K) = 0$ if $\alpha_1 > \alpha$. Thus,

$$\inf\{\alpha: \mu^\alpha(K) = 0\} = \sup\{\alpha: \mu^\alpha(K) = \infty\}$$

and we define the Hausdorff dimension of K as

$$\dim_H(K) = \inf\{\alpha: \mu^\alpha(K) = 0\}.$$

It is known that $\dim(K) \leq \dim_H(K)$ and these numbers are equal when K is a submanifold of a Banach space. For general K, there is little that can be said relating these numbers.

To define another measure of the size of a metric space K, let $N(\varepsilon,K)$ be the minimum number of open balls of radius ε needed to cover K. Define the <u>limit capacity</u> $c(K)$ of K by

$$c(K) = \limsup_{\varepsilon\to 0} \frac{\log N(\varepsilon,K)}{\log (1/\varepsilon)} .$$

In other words, $c(K)$ is the minimum real number such that, for every $\sigma > 0$, there is a $\delta > 0$ such that

$$N(\varepsilon,K) \leq (\tfrac{1}{\varepsilon})^{c(K)+\sigma} \quad \text{if} \quad 0 < \varepsilon < \delta.$$

It is not difficult to show that

$$\dim_H(K) \leq c(K).$$

Another useful property is that, given a Banach space E, a finite dimensional linear subspace S of E with $n = \dim S$, a map $L \in \mathscr{L}(E)$, and using the notation $B_\varepsilon^S(0) = \{v \in S: ||v|| < \varepsilon\}$, we have

$$N(\varepsilon_1,B_{\varepsilon_2}^S(0)) \leq n2^n(1 + \frac{\varepsilon_1}{\varepsilon_2})^n, \quad \text{for all} \quad \varepsilon_1,\varepsilon_2 > 0, \tag{6.1}$$

and

$$N((1+\gamma)\lambda\varepsilon, LB_\varepsilon(0)) \leq n2^n(1 + \frac{||L||+\lambda}{\lambda\gamma})^n \tag{6.2}$$

for all $\gamma,\varepsilon > 0$, $\lambda > ||L_S||$, where $B_\varepsilon(0) = B_\varepsilon^E(0)$ and $L_S\colon E/S \to E/L(S)$ is the linear map induced by S.

Estimates for the limit capacity of the attractor set $A(F)$ of an RFDE will be obtained by an application of general results for the capacity of compact subsets of a Banach space E with the property that $f(K) \supset K$ for some C^1 map $f\colon U \to E$, $U \supset K$, whose derivative can be decomposed as a sum of a compact map and a contraction.

We begin with some notation. For $\lambda > 0$, the subspace of $\mathscr{L}(E)$ consisting of all maps $L = L_1 + L_2$ with L_1 compact and $||L_2|| < \lambda$ is denoted by $\mathscr{L}_\lambda(E)$. Given a map $L \in \mathscr{L}_\lambda(E)$ we define

$$\nu_\lambda(L) = \min\{\dim S\colon S \text{ is a linear subspace of } E \text{ and } ||L_S|| < \lambda\}.$$

It is easy to prove that $\nu_\lambda(L)$ is finite for $L \in \mathscr{L}_{\lambda/2}(E)$.

<u>Theorem 6.1</u>. <u>Let</u> E <u>be a Banach space</u>, $U \subset E$ <u>an open set</u>, $f\colon U \to E$ <u>a</u> C^1 <u>map</u>, <u>and</u> $K \subset U$ <u>a compact set such that</u> $f(K) \supset K$.

<u>If the Fréchet derivative</u> $D_xf \in \mathscr{L}_{1/4}(E)$ <u>for all</u> $x \in K$, <u>then</u>

$$c(K) \le \frac{\log\{\nu[2(\lambda(1+\sigma) + k^2)/\lambda\sigma]^\nu\}}{\log[1/2\ \lambda(1+\sigma)]} \tag{6.3}$$

<u>where</u> $k = \sup_{x\in K}||D_xf||$, $0 < \lambda < 1/2$, $0 < \sigma < (1/2\lambda)-1$, $\nu = \sup_{x\in K} \nu_\lambda(D_xf^2)$.

<u>If</u> $D_xf \in \mathscr{L}_1(E)$ <u>for all</u> $x \in K$, <u>then</u> $c(K) < \infty$.

<u>Proof</u>: Assume that $D_xf \in \mathscr{L}_{1/4}(E)$, $x \in K$. Then for some $0 \le \lambda \le 1/8$, $D_xf^2 \in \mathscr{L}_{\lambda/2}(E)$ for all $x \in K$. By the remark just preceding this theorem, for each $x \in K$, there exists a finite dimensional linear subspace $S(x)$ of E such that $||(D_xf^2)S(x)|| < \lambda$, and, by continuity, $||(D_yf^2)S(x)|| < \lambda$

for every y in some neighborhood of x. We construct in this way an open covering of K which can be taken finite, since K is compact. It follows that $\nu = \sup_{x \in K} \nu_\lambda(D_x f^2) < \infty$. Take $\delta > 1$ and $\sigma > 0$ satisfying $(1+\sigma)\lambda\delta < 1/2$. By the continuity of f^2, there exists $\varepsilon_0 > 0$ such that $f^2 B_\varepsilon(x) \subset f^2(x) + (D_x f^2) B_{\delta\varepsilon}(0)$ for all $x \in K$, $0 < \varepsilon < \varepsilon_0$. Without loss of generality, we can take $\varepsilon_0 < 1$.

Let $\lambda_0 = (1+\sigma)\delta\lambda$, and $\lambda_1 = \nu 2^\nu (1 + \frac{k^2+\lambda}{\lambda\sigma})^\nu$, where k is as in the statement of the theorem. Then, since $||D_x f^2|| \leq k^2$, the inequality (6.2) gives

$$N(\lambda_0\varepsilon, f^2 B_\varepsilon(x)) \leq N(\lambda_0\varepsilon, (D_x f^2) B_{\delta\varepsilon}(0)) \leq \lambda_1$$

for all $0 < \varepsilon < \varepsilon_0$. Since K is compact, it can be covered by a finite number of balls $B_\varepsilon(x_i)$, $x_i \in K$. It follows that $K \subset f(K) \subset f^2(K) \subset \bigcup_i f^2 B_\varepsilon(x_i)$. Therefore, the last inequality implies

$$N(\lambda_0\varepsilon, K) \leq \lambda_1 N(\varepsilon, K) \leq \lambda_1 N(\varepsilon/2, K),$$

for all $0 < \varepsilon < \varepsilon_0$. Since $2\lambda_0 < 1$, each ε in the interval $0 < \varepsilon < \lambda_0\varepsilon_0$ can be written as $\varepsilon = (2\lambda_0)^p \bar{\varepsilon}$ for some $\bar{\varepsilon}$ in the interval $\lambda_0\varepsilon_0 < \bar{\varepsilon} < \varepsilon_0/2$ and some integer $p \geq 1$, and, therefore, the last inequality can be applied p times to get

$$N(\varepsilon, K) = N((2\lambda_0)^p\bar{\varepsilon}, K) \leq \lambda_1^p N(\bar{\varepsilon}, K) \leq \lambda_1^p N(\lambda_0\varepsilon_0, K).$$

Writing $\varepsilon_1 = \lambda_0\varepsilon_0$, we have

$$\frac{\log N(\varepsilon,K)}{\log (1/\varepsilon)} \le \frac{p \log \lambda_1 + \log N(\varepsilon_1,K)}{p \log (1/2\lambda_0)} \le \frac{\log \lambda_1}{\log(1/2\lambda_0)} + \frac{\log N(\varepsilon_1,K)}{\log(1/2\lambda_0)} \cdot \frac{\log 2\lambda_0}{\log(\varepsilon/\lambda_0\varepsilon_0)} .$$

Taking the lim sup as $\varepsilon \to 0$, we obtain

$$c(K) \le \frac{\log \lambda_1}{\log(1/2\lambda_0)} .$$

Since this inequality holds for any $\delta > 1$ and $\lambda_0 = (1+\sigma)\delta\lambda$, we get

$$c(K) \le \frac{\log \lambda_1}{\log(1/2(1+\sigma)\lambda)}$$

which is precisely the inequality (6.3) in the first statement in the theorem.

In order to prove the second statement in the theorem, one just notes that, if $D_x f \in \mathscr{L}_1(E)$ for all $x \in K$, then the continuity of $D_x f$ and the compactness of K imply the existence of $0 < \lambda < 1$ such that $D_x f \in \mathscr{L}_\lambda(E)$ for all $x \in K$. Consequently, for every integer $p \ge 1$, $D_x f^p \in \mathscr{L}_{\lambda^p}(E)$ for all $x \in K_p = \bigcap_{j=0}^{p} f^{-j}(K)$. Taking p sufficiently large for $\lambda^p < 1/4$, the first statement of the theorem implies $c(K_p) < \infty$. But $K_p \subset K \subset f^p(K_p)$ implies $c(K_p) \le c(K) \le c(f^p K_p)$, and, since f^p is a C^1 map, it does not increase the capacity of compact sets. Therefore, $c(K) = c(K_p)$, and the proof of the theorem is complete.

Theorem 6.2. *Let* $F \in \mathscr{X}^1$ *be an RFDE on a manifold* M, *and* $A_\beta(F) = A(F) \cap \{\varphi \in C^0: |\varphi| \leq \beta\}$. *There is an integer* d_β, *depending only on* M, *the delay* r *and the norm of* F, *such that*

$$c(A_\beta(F)) \leq d_\beta \text{ for all } \beta \in [0,\infty).$$

Consequently, also $\dim_H A_\beta(F) \leq d_\beta$, $\beta \in [0,\infty)$ and $\dim_H A(F) < \infty$ when M is compact.

Proof: The case of noneuclidean manifolds M can be reduced to the case of an RFDE defined on $\mathbb{R}^k$, for an appropriate integer k, by the Whitney imbedding theorem and considering an RFDE on $\mathbb{R}^k$ defined by an extension of F to $\mathbb{R}^k$ similar to the one constructed in the proof of Theorem 2.1. Consequently, we take without loss of generality $M = \mathbb{R}^m$.

The Ascoli Theorem guarantees that $A_\beta(F)$ is compact, and consequently we can take a bounded open neighborhood $U \supset A_\beta(F)$, such that $\Phi_r U$ is precompact. It can be easily shown that $D_x\Phi_r$ is a compact operator for each $x \in U$. On the other hand, from the definition of $A_\beta(F)$ it follows $\Phi_r(A_\beta(F)) \supset A_\beta(F)$. Consequently, we can apply Theorem 6.1 with $K = A_\beta(F)$ and $f = \Phi_r$, $k = \sup_{\varphi \in A_\beta(F)} ||D_\varphi \Phi_r||$, while taking $\sigma = 1$ and $0 < \lambda < \min(k/4, 1/4)$, to get

$$c(A_\beta(F)) \leq \frac{\log\{\nu[(4\lambda+2k^2)/\lambda]^\nu\}}{\log[1/4\lambda]} < \infty,$$

The bound of $\dim_H A_\beta(F)$ follows immediately and then it is clear that $\dim_H A(F) < \infty$ when M is compact.

Another result guarantees that the attractor set $A(F)$ can be "flattened" by any projection of a residual set of projections π from C^0 into a finite dimensional linear subspace of C^0 with sufficiently high dimension, in the sense that the restriction of π to $A(F)$ is one-to-one. This result is included here because it is of possible importance for the study of A-stability and bifurcation. It uses the following:

Theorem 6.3. If E is a Banach space and $A \subset E$ is a countable union of compact subsets K_i of E and there exists a constant D such that $\dim_H(K_i \times K_i) < D$ for all i, then, for every subspace $S \subset E$ with $D + 1 < \dim S < \infty$, there is a residual set $\mathscr{R}$ of the space $\mathscr{P}$ of all continuous projections of E onto S (taken with the uniform operator topology) such that the restriction π/A is one-to-one for every $\pi \in \mathscr{R}$.

Proof: We transcribe the proof given by Mañé. Suppose $A = \bigcup_{i=1}^{\infty} K_i$ where each K_i is compact, and take S and $\mathscr{P}$ to be as in the statement of this theorem. Denote

$$P_{i,\varepsilon} = \{\pi \in \mathscr{P}: \operatorname{diam}(\pi^{-1}(p) \cap K_i) < \varepsilon \quad \text{for all} \quad p \in S\},$$

where diam denotes the usual diameter of a set in a metric space. Clearly, $P_{i,\varepsilon}$ is open and $\mathscr{R} = \bigcap_{i,j=1}^{\infty} P_{i,1/j}$ is the set of projections onto S which are one-to-one when restricted to A. It is, therefore, sufficient to prove that every $P_{i,\varepsilon}$ is dense in $\mathscr{P}$.

Let $Q_{i,\varepsilon} = \{v-w: v,w \in K_i$ and $||v-w|| \geq \varepsilon\}$. Then $\pi \in P_{i,\varepsilon}$ if and only if $\pi^{-1}(0) \cap Q_{i,\varepsilon} = \emptyset$. Denote by h the canonical homomorphism of E onto the quotient space E/S. The set $h(Q_{i,\varepsilon}) - \{0\}$ is a countable union of the compact sets $\mathscr{I}_j = \{h(v): v \in Q_{i,\varepsilon}, ||h(v)|| \geq 1/j\}$, $j = 1,2,\ldots$. Therefore there exists a sequence $L_k: E/S \to \mathbb{R}$ of continuous linear maps such that $L_k(x) = 0$ for all $k \in \mathbb{N}$, implies $x \notin h(Q_{i,\varepsilon}) - \{0\}$. Let

$$Q_{i,\varepsilon,k,j} = \{v \in Q_{i,\varepsilon}: |L_k(h(v))| \geq 1/j\}$$

and

$$P_{i,\varepsilon,k,j} = \{\pi \in \mathscr{P}: \pi^{-1}(0) \cap Q_{i,\varepsilon,k,j} = \emptyset\}.$$

Then each $P_{i,\varepsilon,k,j}$ is open and $P_{i,\varepsilon} = \bigcap_{k,j} P_{i,\varepsilon,k,j}$. Consequently, it is sufficient to show that every $P_{i,\varepsilon,k,j}$ is dense in $\mathscr{P}$.

Let $\pi_0 \in \mathscr{P}$, $C = \{v \in S: ||v|| = 1\}$ and define $\zeta: S - \{0\} \to C$ by $\zeta(v) = v/||v||$. Then

$$\dim_H \zeta(\pi_0 Q_{i,\varepsilon}) \leq \sup_{\delta>0} \dim_H \zeta[(\pi_0 Q_{i,\varepsilon}) \cap (S-B_\delta(0))].$$

But the restriction of ζ to $[(\pi_0 Q_{i,\varepsilon}) \cap (S-B_\delta(0))]$ is Lipshitz and, therefore,

$$\dim_H \zeta[(\pi_0 Q_{i,\varepsilon}) \cap (S-B_\delta(0))] \leq \dim_H \pi_0 Q_{i,\varepsilon}.$$

It follows that $\dim_H \zeta(\pi_0 Q_{i,\varepsilon}) \leq \dim_H Q_{i,\varepsilon} \leq \dim_H (K_i \times K_i)$. Since $\dim_H C = \dim S-1 > \dim_H (K_i \times K_i)$, there exists $u \in C$ such that $u \notin \zeta(\pi_0 Q_{i,\varepsilon})$. Given $\delta > 0$ and an integer k, let us consider $\pi_{\delta,k} \in \mathscr{P}$ given by $\pi_{\delta,k} = \pi_0 + \delta u L_k \circ h$. Assume $\pi_{\delta,k}(x) = 0$ and $x \in Q_{i,\varepsilon,k,j}$ then $\pi_0(x) = -\delta L_k(h(x))u$ and $L_k(h(x)) \neq 0$ and $\pi_0(x) \neq 0$. Consequently,

$$u = -[\delta L_k(h(x))]^{-1}\pi_0(x)$$

and, then, $u = \zeta(u) = \zeta(\pi_0(x)) \in \zeta(\pi_0 Q_{i,\varepsilon})$, contradicting the choice of u. This proves that $\pi_{\delta,k}^{-1}(0) \cap Q_{i,\varepsilon,k,j} = \emptyset$ and, therefore, $\pi_{\delta,k} \in P_{i,\varepsilon,k,j}$. Since $\pi_{\delta,k} \to \pi_0$ as $\delta \to 0$, this proves $\mathscr{P}_{i,\varepsilon,k,j}$ is dense in $\mathscr{P}$.

<u>Theorem 6.4</u>. <u>Let</u> $F \in \mathscr{X}^1$ <u>be an RFDE on</u> $\mathbb{R}^m$. <u>There is an integer</u> d, <u>depending only on</u> m, <u>the delay</u> r <u>and the norm of</u> F, <u>such that, if</u> S <u>is a linear subspace of</u> C^0 <u>with</u> $d \leq \dim S < \infty$, <u>then there is a residual set</u> $\mathscr{R}$ <u>of the space of all continuous projections of</u> C^0 <u>onto</u> S, <u>such that the restriction</u> $\pi/A(F)$ <u>is one-to-one for every</u> $\pi \in \mathscr{R}$.

<u>Proof</u>: Apply Theorem 6.3 with $E = C^0$, $A = A(F) = \bigcup_{\beta=1}^{\infty} A_\beta(F)$, taking into account Theorem 6.2 and the fact that $A_\beta(F)$ is compact for every $\beta > 0$.

If M is a compact manifold, it is possible to obtain more information on the dimension of the attractor set using algebraic topology.

<u>Lemma 6.5</u>. <u>Suppose</u> M <u>is a compact manifold</u>. <u>Then the map</u> $\rho|A(F)$: $\varphi \in A(F) \mapsto \varphi(0)$, <u>induces an injection</u> $(\rho|A(F))^*: H^*(M) \to H^*(A(F))$ <u>on Čech cohomology</u>.

<u>Proof</u>: Define $\psi: M \to C^0$ by $\psi(p)(t) = p$ for all $-r \leq t \leq 0$. Then $\rho\psi$ is the identity and $\psi\rho$ is homotopic to the identity. Therefore, $\rho^*: H^*(M) \to H^*(C^0)$, the induced map on Čech cohomology, is the identity. But, if $i: A(F) \to C^0([-r,0],M)$ denotes the inclusion map, we have $(\rho/A(F))^* = (\rho i)^* = i^*\rho^*$. Thus, we have reduced the problem to the injectivity of i^* which, by the continuity property of Čech cohomology, is

reduced to showing that if $K_n = C\ell\, \Phi_r^n(C^0)$ and $i_n: K_n \to C^0$ is the inclusion map then i_n^* is injective for all n (recall that $\bigcap_{n\geq 0} K_n = A(F)$). But we can write $\Phi_r^n = i_n g_n$, with $g_n: C^0 \to K_n$, and then $\Phi_r^{*n} = g_n^* i_n^*$. Now observe that if $\Phi_t^{(\lambda)}$, $t \geq 0$, is the solution map on C^0 defined by the RFDE $\dot{x}(t) = \lambda\Gamma(x_t)$, then $\Phi_r^{(1)} = \Phi_r$, $\Phi_r^{(0)} = \psi\rho$ and the maps $\psi\rho$ and Φ_r are homotopic. Hence $g_n^* i_n^* = \Phi_T^{*n} = (\psi\rho)^{*n} = I$ and i_n^* is injective.

A consequence of Lemma 6.5 is the following

<u>Theorem 6.6</u>. <u>Let</u> $F \in \mathscr{X}^1$ <u>be an RFDE on a compact manifold</u> M. <u>Then</u> $\dim A(F) \geq \dim M$, <u>and the map</u> $\rho: \varphi \to \varphi(0)$ <u>maps</u> $A(F)$ <u>onto</u> M, <u>that is, through each point of</u> M <u>passes a global solution</u>.

<u>Proof</u>: Let $m = \dim M$. Since $H^{*k}(M) = 0$ for $k < m$, $H^{*m}(M)$ is nontrivial and $A(F)$ is compact by Corollary 5.4, the first and last statements of the theorem follow from the preceding lemma. Suppose ρ does not take $A(F)$ onto M. Then there is a p in M such that $\rho(A(F)) \subseteq M\setminus\{p\}$. But $H^m(M) = \mathbb{Z}_2$ and $H^m(M\setminus\{p\}) = 0$, which is a contradiction.

The following example shows how easily one can construct examples where the evaluation map ρ is not one-to-one on $A(F)$.

<u>Example</u>. Let S^1 be the circle. Any point p in S^1 is determined by an angle x and given p, x is only determined up to a multiple of 2Π. The unit tangent vector u_p at the point p is equal to

$$u_p = -(\sin x)\cdot\vec{i} + (\cos x)\cdot\vec{j}$$

The function $g: S^1 \to R$ given by

$$g(p) = \frac{\pi}{2}(1 - \cos x)$$

defines an RFDE on S^1 in the following way:

$$\varphi \in C^0(I,S^1) \mapsto f(\varphi) = [g(\varphi(0)) + g(\varphi(-1))]\cdot u_{\varphi(0)}.$$

A solution $p(t)$ satisfies $\dot{p}(t) = f(p_t)$ where $p(t) = 0 + \cos x(t)\vec{i} + \sin x(t)\vec{j}$ and then

$$\dot{p}(t) = (-\sin x(t)\vec{i} + \cos\cdot x(t)\vec{j})\cdot x(t) = [g(p(t)) + g(p(t-1))]u_{p(t)}$$

or

$$\dot{x}(t) = \frac{\pi}{2}(1 - \cos x(t)) + \frac{\pi}{2}(1 - \cos x(t-1)) \tag{6.4}$$

The constant solutions of (6.4) must satisfy $2 = \cos x(t) + \cos x(t-1)$. Thus $x(t) = 2k\pi$, $k = 0,\pm1,\pm2,\ldots$. The only critical point is $P = 0 + \vec{i}$. On the other hand, $x(t) = \pi t$ is a solution of (6.4) and on S^1 the corresponding periodic solution is given by $p(t) = 0 + (\cos \pi t)\vec{i} + (\sin \pi t)\vec{j}$. Thus, ρ is not one-to-one on $A(F)$.

Theorem 6.6 does not hold when M is not compact. Consider in $M = \mathbb{R}^3$ the system

$$\begin{aligned} \dot{x}(t) &= 2y(t) \\ \dot{y}(t) &= -z(t) + x(t-1) \\ \dot{z}(t) &= 2y(t-1) \end{aligned} \tag{6.5}$$

A simple computation shows that for $t \geq 1$ one obtains $\ddot{y}(t) = \dddot{x}(t) = 0$ and any solution $(x(t),y(t),z(t))$ must lie in the plane

$$x(t) - 2y(t) - z(t) = 0.$$

The finite-dimensionality of the sets $A_\beta(F)$ implies the finite-dimensionality of the period module of any almost periodic solution of F, generalizing what happens for ordinary differential equations. Let us recall the definition of period module. Any almost periodic function $x(t)$ has a Fourier expansion

$$x(t) \sim \sum a_n e^{-i\lambda_n t}$$

where $\Sigma|a_n|^2 < \infty$; the period module of $x(t)$ is the vector space $\mathcal{M}$ spanned by $\{\lambda_n\}$ over the rationals. The fact that the period module is finite dimensional implies that the almost periodic solution is quasiperiodic.

Corollary 6.7. Let $F \in \mathcal{X}^1$ be an RFDE on a manifold M. Then there is an integer N depending only on the delay r, the norm of F and on M such that, for any almost periodic solution $x(t)$ of F, the period module $\mathcal{M}$ of x has finite-dimension $\leq N$; that is, there are only finitely many rationally independent frequencies in the Fourier expansion for x.

Proof: An easy modification of a result given by Cartwright for ordinary differential equations shows that $\dim \mathcal{M}$ equals the topological dimension of the hull $\mathcal{H}$ of x. Clearly, $\mathcal{H}$ is homeomorphic to the set of initial data at $t = 0$ for its elements, which is a subset of $A_\beta(F)$, where β is a bound on the solution x. Since $A_\beta(F)$ is finite-dimensional by Theorem 6.1, so are $\mathcal{H}$ and $\mathcal{M}$.

The set $A(f)$ may not have finite dimension if f is in $\mathscr{X}^0$. In fact, let Q_L be the set of functions $\gamma: \mathbb{R}^n \to \mathbb{R}^n$ with global Lipschitz constant L. For each $\gamma \in Q_L$, each solution of $\dot{x}(t) = \gamma(x(t))$ is defined for all $t \in \mathbb{R}$. One can prove the following result.

Theorem 6.8. *For each $L > 0$ there is a continuous RFDE(f) on $\mathbb{R}^n$, depending only on L, such that, for every $\gamma \in Q_L$, every solution of $\dot{x}(t) = \gamma(x(t))$ is also a solution of the RFDE(f). In particular, $A(f)$ has infinite dimension.*

7. Attractor Sets as C^1-Manifolds

It is of some interest to determine when the attractor $A(F)$ is a C^1-manifold, since it will then have a particularly simple geometric structure which will facilitate the study of qualitative properties of the flow. Results in this direction can be established through the use of C^k-*retractions* which are defined as C^k maps γ from a Banach manifold into itself such that $\gamma^2 = \gamma$, $k \geq 1$.

Lemma 7.1. If B *is a Banach manifold (without boundary) and* $\gamma: B \to B$ *is a* C^1*-retraction, then* $\gamma(B)$ *is a Banach* C^1*-submanifold of* B *(without boundary).*

Proof: Since $\gamma \cdot \gamma = \gamma$, the derivative T of γ at a point $p \in \gamma(B)$, $T = \gamma'_p$, satisfies $\gamma'_{\gamma(p)} \cdot \gamma'_p = \gamma'_p$, that is, $T^2 = T$. This implies $T = \gamma'_p$ is double splitting; in fact, taking $E = T_pB$, $T: E \to E$, then the image and kernel of T are, respectively, the kernel and image of $(I - T)$. The local representative theorem shows that with suitable local charts called there α and β, the map γ can be represented by

$$\bar{\gamma}: (u,v) \to (u,\eta(u,v)),$$

for $(u,v) \in B_1 \times B_2$ where B_1 and B_2 are the open unit balls in E_1 = image of T and E_2 = kernel of T and, the map $\eta(u,v)$ satisfies $D\eta(0,0) = 0$. Consider the points $(u,v) \in B_1 \times B_2$ such that $\eta(u,v) = v$. By the implicit function theorem, there exist open balls $\bar{B}_1 \subseteq B_1$ and $\bar{B}_2 \subseteq B_2$ such that the set of points in $\bar{B}_1 \times \bar{B}_2$ satisfying $\eta(u,v) = v$ is the graph of a function $v = \bar{v}(u)$, $u \in \bar{B}_1$. The Banach manifold $\bar{M}$ defined by that graph in B is locally contained in the image of $\bar{\gamma}$ since

$\gamma(u,\bar{v}(u)) = (u,\eta(u,\bar{v}(u))) = (u,\bar{v}(u))$. On the other hand, the map γ can be also represented by

$$\bar{\bar{\gamma}} = \alpha\cdot\gamma\cdot\alpha^{-1}: \quad (u,v) \to (f(u,v),g(u,v))$$

and the partial derivative $D_v g(0,0)$ is zero since $v \in E_2$ = kernel of T. The equation $g(u,v) = v$ has a local solution $v = \bar{\bar{v}}(u)$ in an open neighborhood denoted again by $\bar{B}_1 \times \bar{B}_2$ and the corresponding graph contains the fixed points of $\bar{\bar{\gamma}}$ which are given by $g(u,v) = v$ and $f(u,v) = u$. The set of fixed points of $\bar{\bar{\gamma}}$ restricted to $\bar{B}_1 \times \bar{B}_2$ is the image of $\bar{\bar{\gamma}}$, since $\bar{\bar{\gamma}}$ is a retraction. Therefore, the graph of $v = \bar{\bar{v}}(u)$ defines in B another Banach manifold $\bar{\bar{M}}$ locally containing the image of γ. Since, locally, $\bar{M} \subseteq$ Image of $\gamma \subseteq \bar{\bar{M}}$ and both are E_1 - Banach manifolds containing the point $p \in \gamma(B)$, we get, locally, $\bar{M}$ = Image of $\gamma = \bar{\bar{M}}$.

In Example 3.2 a C^1 vector field X defined on a manifold M was used to define an RFDE on M by $F = X\circ\rho$. The map $\Sigma_X: M \to C^0(I,M)$ such that $\Sigma_X(p)$ is the restriction of the solution of X, through p at $t = 0$, to the interval $I = [-r,0]$, is a cross-section with respect to ρ and $A(F) = \Sigma_X(M)$. The map $\gamma = \Sigma_X\cdot\rho$ is a C^1-retraction and commutes with the flow of F, in agreement with $A(F)$ being a C^1-manifold diffeomorphic to M and invariant under F.

Theorem 7.2. Let $F \in \mathscr{X}^1$ be an RFDE on a compact and connected manifold M and assume there exists a C^1-retraction $\gamma: C^0 \to C^0$ such that $A(F) = \gamma(C^0)$. Then, the attractor set $A(F)$ is a connected compact C^1-manifold. Besides, if γ is homotopic to the identity, $A(F)$ is diffeomorphic to M.

Proof: We know that $A(F)$ is a connected and compact set; by Lemma 7.1 $A(F)$ is a C^1-manifold without boundary. Arguments like the ones used in Lemma 6.5 and Theorem 6.6 show that $A(f)$ is diffeomorphic to M.

Theorem 7.3. Let $F \in \mathscr{X}^1$ be an RFDE on a compact and connected manifold M without boundary, and assume there is a constant $k > 0$ such that $||d\Phi_t(\varphi)|| \leq k$ and $d\Phi_t$ has Lipschitz constant k for all $t \geq 0$ and $\varphi \in C^0$. Then, each point of the attractor set $A(F)$ is an ω-limit point of some point of $A(F)$, and there exists one unique C^1-retraction γ of C^0 onto $A(F)$ which commutes with the flow, i.e., $\gamma\Phi_t = \Phi_t\gamma$, $t \geq 0$. The attractor set $A(F)$ is a connected compact C^1-manifold without boundary and the restriction of Φ_t, $t \geq 0$, to $A(F)$ is a one-parameter group of diffeomorphisms.

Proof: Let t_n be a sequence of real numbers such that $t_n \to \infty$ and $s_n = (t_n - t_{n-1}) \to \infty$. Since M is bounded and $F \in \mathscr{X}^1$, the set $K = \Phi_r(C^0)$ is precompact. For n large enough, the restrictions of Φ_{t_n-r} to K belong to a set of equicontinuous functions and, for each $\varphi \in K$, the set of all $\Phi_{t_n-r}(\varphi)$ is relatively compact. Then by Ascoli's theorem, for a subsequence, denoted again by t_n, the Φ_{t_n-r} converge to a continuous map $\overline{\beta}\colon K \to C$, uniformly on K. It follows easily that Φ_{t_n} converges to the map $\beta = \overline{\beta}\cdot\Phi_r$, uniformly on the Banach manifold $C^0(I,M)$. Using the same argument, there is a subsequence of Φ_{s_n} which converges to a map γ, uniformly on $C^0(I,M)$. Then γ is continuous and Lipschitz with constant k since $||d\Phi_t||$ is bounded by k for all $t \geq 0$. For $\varphi \in C^0$ we have $\beta(\varphi) \in \omega(\varphi)$ and $\omega(\varphi) \subseteq A(F)$. On the other hand, given $\overline{\varphi} \in A(F)$ and $t_n \in \mathbb{R}$, there exist $\Psi_n \in A(F)$ such that $\overline{\varphi} = \Phi_{t_n}(\Psi_n)$. When $n \to \infty$, there is a subsequence (denoted with the same indices) such that $\Psi_n \to \Psi$ and $\Psi \in A(F)$ because $A(F)$

is invariant and closed. Therefore, locally and for n large enough,

$$||\Phi_{t_n}(\Psi_n) - \beta(\Psi)|| \le ||\Phi_{t_n}(\Psi_n) - \Phi_{t_n}(\Psi)|| + ||\Phi_{t_n}(\Psi) - \beta(\Psi)||$$

$$\le k||\Psi_n - \Psi|| + ||\Phi_{t_n}(\Psi) - \beta(\Psi)|| < \varepsilon,$$

which implies $\beta(\Psi) = \overline{\varphi}$ and, consequently, $\beta: C^0 \to A(F)$ is onto. Since $\beta(\Psi) \in \omega(\varphi)$ and $\psi \in A(F)$, it follows that each point of $A(F)$ is in the ω-limit set of some point of $A(F)$, as stated in the theorem.

Now, the relations $\Phi_{s_n} \cdot \Phi_{t_{n-1}} = \Phi_{t_n} = \Phi_{t_{n-1}} \cdot \Phi_{s_n}$ show that $\gamma\cdot\beta = \beta = \beta\cdot\gamma$. Also $\gamma(C^0) = A(F)$ and, then, the map $\gamma: C^0 \to A(F)$ is a retraction since, for any $\overline{\varphi} \in A(F)$, there exists Ψ such that $\beta(\Psi) = \overline{\varphi}$ which implies $\gamma(\overline{\varphi}) = \gamma(\beta(\Psi)) = \beta(\Psi) = \overline{\varphi}$. Finally γ commutes with Φ_t, $t \ge 0$ since

$$\Phi_t \cdot [\gamma(\varphi)] = \Phi_t(\lim \Phi_{s_n}(\varphi) = \lim[\Phi_t \cdot \Phi_{s_n}(\varphi)]$$

$$= \lim[\Phi_{s_n}\Phi_t(\varphi)] = \gamma[\Phi_{t_n}(\varphi)].$$

If $\overline{\gamma}$ is another retraction onto $A(F)$ and $\overline{\gamma}\cdot\Phi_t = \Phi_t\cdot\overline{\gamma}$, $t \ge 0$, then $\overline{\gamma}\cdot\gamma = \gamma\cdot\overline{\gamma}$. Now $\gamma(\varphi) = \overline{\gamma}(\gamma(\varphi)) = \gamma(\overline{\gamma}(\varphi)) = \overline{\gamma}(\varphi)$, i.e. $\gamma = \overline{\gamma}$. This proves uniqueness.

We need to show that γ is C^1. Denote $M_1 = \{v \in TM: ||v|| \le 1\}$ and $C^0(I,M_1) = \overline{M}_1$, the set of all $\psi \in C^0(I,TM)$ such that $||\psi|| \le 1$. Let $\Psi_t = T\Phi_t$ be the flow on $TC^0(I,M)$ of the first variational equation. The set $\Psi_r\overline{M}_1$ is relatively compact and its closure is a compact set K_1; this follows from the boundedness of $||d\Phi_t(\varphi)||$ and $F \in \mathscr{X}^1$.

Consider now the sequence of functions $\Psi_{s_n - r}: K_1 \to TC^0$ which are equicontinuous since $d\Phi_t$ has Lipschitz constant k. Then for each $(\varphi,\psi) \in K_1$, the set of all

$\Psi_{s_n - r}(\varphi,\psi)$ is relatively compact. Thus, there is a subsequence of $(\Psi_{s_n - r})$ which converges uniformly on K_1 and, therefore, (Ψ_{s_n}) converges uniformly on $\overline{M}_1$ to a map $\overline{\gamma}$ which must be the derivative of γ and, consequently, γ is C^1.

Now, Theorem 7.2 implies all the other statements in the theorem, except that Φ_t is a group of diffeomorphisms on $A(F)$, which is, therefore, the only thing that is left to prove.

The solution map $\Phi_t\colon A(F) \to A(F)$ is C^1 differentiable. Let φ and ψ be two elements of $A(F)$. If, for $t = \overline{t}$, $\Phi_{\overline{t}}(\varphi) = \Phi_{\overline{t}}(\psi)$ with $\varphi \neq \psi$ we get $\Phi_t(\varphi) = \Phi_t(\psi)$ for all $t \geq \overline{t}$. Using the sequence Φ_{s_n} which has defined the retraction γ, we get $\Phi_{s_n}(\varphi) = \Phi_{s_n}(\psi)$, and, therefore, $\gamma(\varphi) = \gamma(\psi)$ and $\varphi = \psi$. Since for each t, Φ_t is one-to-one on $A(F)$ and $A(F)$ is compact, Φ_t is a homeomorphism. Also, in the manifold $C^0(I,M_1)$, one has an attractor, defined in a similar way by a retraction, the derivative of γ, obtained by the uniform convergence of the flow Ψ_{s_n} on the manifold $\overline{M}_1 = C^0(I,M_1)$. This shows that the map ψ_t is also one-to-one, and, consequently, Φ_t is a diffeomorphism. Denoting by Φ_{-t} the inverse of Φ_t, $t \geq 0$, one obtains a one-parameter group of C^1-diffeomorphisms acting on the compact manifold $A(F)$.

Remark. The hypothesis $||d\Phi_t(\varphi)|| \leq k$ for all $t \geq 0$ and all $\varphi \in C^0(I,M)$, in Theorem 7.3 is assured by the following geometric condition: the first variational equation restricted to the manifold $\{\psi \in C^0(I,TM) : |\psi| \leq k\}$ is such that its values are vectors tangent to the manifold $\{v \in TM: |v| \leq k\}$, and at points ψ such that $|\psi(0)| = k$, the value of the first variational equation is an "inward" vector.

Example. Consider the RFDE on the circle S^1 given by the scalar equation (see Section 3.10)

$$\dot{x} = b[\sin(x(t) - x(t-1))], \tag{7.1}$$

where $b: R \to R$ is a C^1 function with Lipschitz first derivative satisfying also $b(0) = 0$ and $\left|\frac{db}{dx}\right| \leq \sigma < 1$. The global solutions of this equation are the constant functions. To see this we consider the map

$$T: z(t) \longrightarrow \int_{t-1}^{t} b[\sin(z(u))]du$$

acting in the Banach space of all bounded continuous functions with the sup norm. It is easy to see that T is a contraction and $z(t) \equiv 0$ is its fixed point. On the other hand, if $x(t)$ is a global solution, $[x(t) - x(t-1)]$ is bounded and

$$x(t) - x(t-1) = \int_{t-1}^{t} b[\sin (x(u) - x(u-1))]du$$

which shows that $x(t) - x(t-1) = 0$ and, using the equation, $\dot{x}(t) \equiv 0$ and $x(t) = \text{constant}$. $A(F)$ is in this case a circle in $C^0(I,S^1)$.

Let $x(t)$ be the solution defined by the initial condition φ at $t = 0$. For $n \geq 2$ one has

$$\max_{u\in[n-1,n]} |x(u) - x(u-1)| \leq \sigma \cdot \max_{u\in[n-2,n-1]} |x(u) - x(u-1)| \leq$$

$$\leq \sigma^2 \cdot \max_{u\in[n-3,n-2]} |x(u)-x(u-1)| \leq \dots \leq \sigma^{(n-2)} \max_{u\in[1,2]} |x(u) - x(u-1)|.$$

Then $\lim_{t\to+\infty} |x(t) - x(t-1)| = 0$ and $\lim_{t\to+\infty} \dot{x}(t) = 0$. Since

$$x(t) = \varphi(0) + \int_0^t b[\sin(x(u) - x(u-1))]du,$$

one has

$$|x(t) - \varphi(0)| \le \int_0^t \sigma|x(u) - x(u-1)|du \le K \frac{1}{1-\sigma}$$

for a suitable K. Thus, $x(t)$ is bounded as $t \to +\infty$. Moreover, given $\varepsilon > 0$, there exists $T(\varepsilon)$ such that

$$|x(t) - x(t')| = |\dot{x}(\xi)| < \varepsilon \quad \text{for} \quad t,t' > T(\varepsilon),$$

and the limit of $x(t)$ as $t \to +\infty$ exists.

The flow Φ_t has a limit:

$$\gamma(\varphi) = \lim_{t\to+\infty} \Phi_t(\varphi) = c \quad \text{(constant solution)}.$$

γ is a C^1-retraction, $\gamma^2 = \gamma$, and $\gamma\cdot\Phi_t = \Phi_t\cdot\gamma$. To prove that γ is C^1 with uniform Lipschitz constant, we need to consider the derivative $d\Phi_t$ which is the flow of the first variational equation:

$$\begin{cases} \dot{x} = b[\sin(x(t) - x(t-1))] \\ \dot{y} = (b \circ \sin)'(x(t) - x(t-1))\cdot[y(t) - y(t-1)]. \end{cases}$$

The critical points in this case are the elements of TS^1. It can be proved that

$$d\gamma(\varphi)\Psi = \lim_{t\to+\infty} d\Phi_t(\varphi)\Psi, \qquad \Psi \in TC^0(I,S^1).$$

The retraction γ has $A(F)$ as image and the example satisfies the hypothesis of Theorem 7.3.

The hypothesis of Theorem 7.3 is very restrictive. In fact, as shown in the theorem, the attractor set must consist of points which are in the ω-limit sets of points in the attractor. However, using infinite dimensional analogues on the continuity properties of a certain class of attractors, it is possible to show that the attractor set of small perturbations of equations satisfying the above hypotheses are also C^1-manifolds. For this, we need some more notation.

Let $F \in \mathscr{X}^1$ be an RFDE on a compact manifold M, such that its attractor set satisfies $A(F) = \gamma(C^0)$ for some C^1-retraction $\gamma\colon C^0 \to C^0$. This implies that $A(F)$ is a compact C^1-manifold and there is a tubular neighborhood U of $A(F)$ in $C^0 = C^0(I,M) \subset C^0(I, \mathbb{R}^N)$ (see Lemma 7.1 and Theorem 2.1). For each $R,L > 0$, let $\mathscr{F}^{0,1}(R,L)$ be defined by

$$\mathscr{F}^{0,1}(R,L) = \{s \in C^0(A(F),U)\colon \gamma s = id(A(F)),\ s' \overset{\text{def}}{=} s - id(A(F))$$

satisfies

$$|s'(u)| \le R \quad \text{and} \quad |s'(u) - s'(v)| \le L|u-v|$$
$$\text{for all } u,v \in A(F)\}.$$

It is not difficult to show that $\mathscr{F}^{0,1}(R,L)$ with distance

$$d(s,s_1) = \sup\{|s'(u) - s_1'(u)|\colon u \in A(F)\}$$

is a complete metric space. For $w_0 \in C^1(U,C^0)$ such that:

i) $w_0\gamma = \gamma w_0$ and $w_0|A(F)$ is a diffeomorphism onto $A(F)$,

ii) $||d(w_0|\gamma^{-1}(p))(p)|| \le \xi < 1$ for all $p \in A(F)$,

iii) $||d(w_0|\gamma^{-1}(p))(p)|| \cdot ||d(w_0|A(F))(p)^{-1}|| \le \xi < 1$ for all $p \in A(F)$,

define $\mathscr{L}^1_{w_0}(\bar{R},\bar{L})$, for $\bar{R}, \bar{L} > 0$, to be the set

$$\mathscr{L}^1_{w_0}(\bar{R},\bar{L}) = \{w \in C^1(U,C^0)\colon |w(u)-w_0(u)| \le \bar{R} \text{ and } ||Dw(u)-Dw_0(u)|| \le \bar{L}$$
$$\text{for all } u \in U\}.$$

Lemma 7.4. Let $F \in \mathscr{X}^1$ be an RFDE on a compact manifold M, such that the attractor $A(F) = \gamma(C^0)$ for some C^1-retraction γ and let $w_0 \in C^1(U,C^0)$ satisfy the above conditions i), ii) and iii). If $w \in \mathscr{L}^1_{w_0}(\bar{R},\bar{L})$ for $\bar{R},\bar{L}$ sufficiently small, then there exists a C^1-manifold B_w diffeomorphic to $A(F)$ which is invariant under w, $B_w \to A(F)$ as $w \to w_0$ in the Hausdorff metric, the restriction of w to B_w is a diffeomorphism, and B_w is uniformly asymptotically stable for the discrete flow defined by w^n, $n = 1,2,3,\ldots$.

Proof: Let $w \in \mathscr{L}^1_{w_0}(\bar{R},\bar{L})$. For each $s \in \mathscr{F}^{0,1}(R,L)$, define $H_s: A(F) \to A(F)$ by $H_s = \gamma ws$. For $\bar{R},\bar{L}$ sufficiently small, H_s is close to $H_s^0 \overset{\text{def}}{=} \gamma w_0 s$, and, since w_0 commutes with γ and $\gamma s = id(A(F))$, we have $H_s^0 = w_0 \cdot id(A(F))$, and, consequently, H_s is a $C^{0,1}$-homeomorphism. On the other hand, for $\bar{R},\bar{L}$ sufficiently small, it can be shown after some computations that the map $\mathscr{K}: \mathscr{F}^{0,1} \to \mathscr{F}^{0,1}$ given by $\mathscr{K}(s) = wsH_s^{-1}$ is a contraction, and, therefore has a unique fixed point $\bar{s} = w\bar{s}H_{\bar{s}}^{-1}$. If we define $B_w = \bar{s}(A(F))$, then B_w is a C^1-manifold diffeomorphic to $A(F)$, and invariant under w because $w\bar{s} = \bar{s}H_{\bar{s}}$ implies $wB_w \subset B_w$. Letting $w \to w_0$, we have $\bar{s} \to id(A(F))$, implying that $B_w \to A(F)$. For $\bar{R},\bar{L}$ sufficiently small, w is close to w_0 in the C^1-uniform norm, and, since w_0 is a diffeomorphism on $A(F)$, it follows that w is a diffeomorphism on B_w. It remains to prove the stability of B_w under $\{w^n\}$.

It is easy to see that, for $\bar{R},\bar{L}$ sufficiently small and for each $y \in U$ sufficiently close to B_w, there exists $s \in \mathscr{F}^{0,1}$ such that $y = s\gamma(y)$.

By the definition of the map $\mathscr{K}$, we have $\mathscr{K}^n(s)\gamma w^n s = w^n s$, which implies $w^n y = \mathscr{K}^n(s)\gamma w^n y$. Due to the properties of $\mathscr{F}^{0,1}$ and since $\mathscr{K}$ is a contraction, we have $\mathscr{K}^n(s) \to \bar{s}$ as $n \to \infty$, uniformly in $s \in \mathscr{F}^{0,1}$. As $B_w = \bar{s}(A(F))$, it follows that, for every $\varepsilon > 0$, there exists $N > 0$ such that $n \geq N$ implies $\operatorname{dist}(\mathscr{K}^n(s)\psi, \bar{s}\psi) < \varepsilon$ for all $\psi \in A(F)$, and consequently, also $\operatorname{dist}(w^n y, B^w) < \varepsilon$. Since the first inequality is uniform in s, it follows that the second is uniform in y. This proves that B_w is uniformly asymptotically stable under the flow $\{w^n\}$.

Theorem 7.5. Let $F \in \mathscr{X}^1$ be an RFDE on a compact manifold M. Suppose there is a constant k such that $||d\Phi_t(\varphi)|| \leq k$ and $d\Phi_t$ has Lipschitz constant k, for all $t \geq 0$ and $\varphi \in C^0$. Then, there is a neighborhood V of F in $\mathscr{X}^1$ such that $A(G)$ is diffeomorphic to $A(F)$ for $G \in V$ and $A(G) \to A(F)$ as $G \to F$. In particular, $A(G)$ is a C^1-manifold, and, if M is connected and without boundary, then $A(G)$ is a connected compact C^1-manifold without boundary and the restriction of Φ_t^G, $t \geq 0$ to $A(G)$ is a one-parameter group of diffeomorphisms.

Proof: Let γ be the retraction onto $A(F)$ constructed in the proof of Theorem 7.3 and U a tubular neighborhood of $A(F)$ in C^0. By continuity of the semiflow map $\Phi(t,\varphi,F)$ and the continuity of the semiflow map for the first variational equation, given $\bar{R}, \bar{L} > 0$, there is a neighborhood V of F in $\mathscr{X}^1$ and some $T > 0$ such that $G \in V$ implies $\Phi_T^G \in \mathscr{L}_\gamma^1(\bar{R}, \bar{L})$. Taking $\bar{R}, \bar{L}$ sufficiently small, Lemma 7.4 can be applied with $w_0 = \gamma$ and $w = \Phi_T^G$ to give $B(G) = B_w$ diffeomorphic to $A(F)$, invariant under Φ_T^G, uniformly asymptotically stable for the flow $(\Phi_T^G)^n$, $n = 1,2,\ldots$, with Φ_T^G being a diffeomorphism on $B(G)$ and $B(G) \to A(F)$ as $G \to F$.

For $t \geq 0$ small and G sufficiently close to F, Φ_t^G is a diffeomorphism on $B(G)$, close to the identity, and $\Phi_T^G \Phi_t^G B(G) = \Phi_t^G \Phi_T^G B(G) = \Phi_t^G B(G)$. Therefore $(\Phi_T^G)^n \Phi_t^G B(G) = \Phi_t^G B(G)$ and, then, the uniform asymptotic stability of $B(G)$ under $(\Phi_T^G)^n$ implies $\Phi_t^G B(G) \subset B(G)$. Thus, $B(G)$ is positively invariant under the flow Φ_t^G, $t \geq 0$. To prove that $B(G)$ is invariant for the RFDE(G), we need to extend the flow of G on $B(G)$ to $t < 0$. Let $\varphi \in B(G)$ and consider the curve $s \to (\Phi_s^G)^{-1}\varphi$ defined for the values $s \leq 0$ for which this curve lies in $B(G)$. Fix $s_0 \leq 0$ in the domain of this curve, choose $t_0 > -s_0 + 2r$ and consider the solution curve $t \to \Phi_t^G[(\Phi_{t_0}^G)^{-1}\varphi]$, $t \geq 0$. For $s \in [-2r + s_0, 0]$ and $t = t_0 + s$ we have

$$\Phi_t^G[(\Phi_{t_0}^G)^{-1}\varphi] = (\Phi_{-s}^G)^{-1}\varphi$$

If

$$y(s+\theta) = \Phi_{t_0+s}^G(\Phi_{t_0}^G)^{-1}\varphi(\theta), \quad s \in [-r+s_0, 0,], \quad \theta \in [-r, 0],$$

then

$$\dot{y}(s) = G(y_s).$$

This shows that $B(G)$ is invariant under the RFDE(G). Thus $B(G) \subset A(G)$.

On the other hand, $(\Phi_T^G)^n B(G) \subset B(G) \subset A(G)$ together with the uniform asymptotic stability of $B(G)$, the upper semicontinuity of $A(G)$ in G (Theorem 5.5) and the fact that $B(G) \to A(F)$ as $G \to F$, imply that $A(G) \subset B(G)$. Thus $A(G) = B(G)$ and the rest of the statement follows from Lemma 7.4.

As mentioned before, if $F \in \mathscr{X}^1$ is an RFDE defined by an ordinary differential equation on a manifold M, as in Example 3.2, the attractor set A(F) is also given by a C^1-retraction. Therefore, the preceding ideas can be applied to establish another class of RFDEs whose attractors are C^1-manifolds, namely the RFDEs close to ordinary differential equations.

Theorem 7.6. *Let* X *be a* C^1-*vector field defined on a compact manifold* M. *There is a neighborhood* V *of* $F = X\circ\rho$ *in* $\mathscr{X}^1$ *such that* A(G) *is a* C^1-*manifold diffeomorphic to* M *for* $G \in V$, $A(G) \to A(F)$ *as* $G \to F$ *and the restriction of* Φ_t^G, $t \geq 0$ *to* A(G) *is a one-parameter family of diffeomorphisms.*

Proof: Let $\Sigma_X: M \to C^0$ be the map such that $\Sigma_X(\varphi)$ is the restriction of the solution of X through φ at $t = 0$, to the interval $I = [-r,0]$. The map $\gamma = \Sigma_X\rho$ is a C^1-retraction which commutes with the flow of F, and $A(F) = \gamma(C^0)$. Given arbitrary $\overline{R},\overline{L} > 0$, there is a neighborhood V of F in $\mathscr{X}^1$ such that $\Phi_r^G \in \mathscr{L}^1_{\Phi_r^F}(\overline{R},\overline{L})$. Lemma 7.4 can now be applied with $w_0 = \Phi_r^F$ and $w = \Phi_r^G$. The rest of the proof is identical to the second part of the proof of the preceding theorem.

Remark. One can obtain higher order of smoothness for the manifolds obtained in the preceding results. In fact, the manifold B_w of Lemma 7.4 will be C^k if w_0, w and γ are of class C^k, $k \geq 1$, and the condition (iii) is replaced by

$$\text{iii)}' \quad ||d(w_0|\gamma^{-1}(p))(p)|| \; ||d(w_0|A(F))(p)^{-1}||^k \leq \xi < 1$$
$$\text{for all } p \in A(F).$$

This last condition holds trivially in theorem 7.6, since $||d(w_0|\gamma^{-1}(p))(p)|| = 0$ for all $p \in A(F)$.

For the case when $M = \mathbb{R}^n$ and F is given by an ordinary differential equation, a result somewhat similar to the preceding theorem, was announced by Kurzweil. The proof given here uses considerations different from the above and having some independent interest. The main idea is to look for the manifold of global orbits by finding the ordinary differential equation defining the flow on that manifold. This is accomplished by using a nonlinear variation of constants formula in such a way that one finds the perturbed invariant manifold by finding first the dynamics on it.

Theorem 7.7. Let $f: \mathbb{R}^n \to \mathbb{R}^n$ be a C^2 function which is bounded and has bounded derivatives, and define $F: C^0(I, \mathbb{R}^n) \to \mathbb{R}^n$ by $F(\varphi) = f(\varphi(0))$. For $G \in \mathscr{X}^1(C^0(I, \mathbb{R}^n), \mathbb{R}^n)$, consider the RFDE given by

$$\dot{x}(t) = G(x_t). \tag{7.1}$$

There exists a neighborhood V of F in $\mathscr{X}^1(C^0(I, \mathbb{R}^n), \mathbb{R}^n)$ such that, for $G \in V$, the set $B(G)$ of all points belonging to orbits of global solutions of (7.1) is diffeomorphic to $\mathbb{R}^n$, depends continuously on G, and the flow of (7.1) in $B(G)$ is given by a one-parameter group of diffeomorphisms, i.e., there exists $g: \mathbb{R}^n \to \mathbb{R}^n$ such that the solutions of $\dot{x}(t) = g(x(t))$ and the global solutions of (7.1) coincide.

Proof: Let us denote by $\xi(t;a,g)$ the value at t of the solution of the ODE $\dot{x}(t) = g(x(t))$ which satisfies the initial condition $x(0) = a$, and set $H(t,a) = \frac{\partial \xi}{\partial a}(t;a,f)$. We can write

$$B(F) = \{\xi(\cdot;a,f): a \in \mathbb{R}^n\}.$$

On the other hand, equation (7.1) can be written as

$$\dot{x}(t) = f(x(t)) + [G(x_t) - f(x(t))].$$

Let $\mathcal{S}^{0,1}(L)$ denote the set

$$\mathcal{S}^{0,1}(L) = \{s \in \mathcal{L}^1(\mathbb{R}^n, C^0): |s(a)| \leq L, \ |s(a)-s(b)| \leq L|a-b| \ \text{for all} \ a,b \in \mathbb{R}^n\}.$$

This set is a complete metric space with distance $d(s,s') = \sup\{|s(a)-s'(a)|: a \in \mathbb{R}^n\}$.

We consider the function defined in $\mathcal{L}^1(\mathbb{R}^n, C^0) \times \mathcal{L}^1(C^0, \mathbb{R}^n)$ by

$$\mathcal{A}(s,G)(a,\theta) = \int_0^\theta H[\theta-\tau, \xi(\tau;a,f) + s(a)(\tau)] \cdot$$

$$\cdot\{G[\xi(\cdot;\xi(\tau;a,f) + s(a)(\tau),f) + s(\xi(\tau;a,f) + s(a)(\tau))(\cdot),f]\}$$

$$- f[\xi(\tau;a,f) + s(a)(\tau),f)]\}d\tau.$$

It can be shown after some computations that, if V is a sufficiently small neighborhood of F in $\mathcal{L}^1(C^0(I, \mathbb{R}^n), \mathbb{R}^n)$, then $\mathcal{A}(\cdot,G)$ is a uniform contraction from $\mathcal{S}^{0,1}(L)$ into $\mathcal{S}^{0,1}(L)$, for $G \in V$. Applying the uniform contraction principle we obtain, for each $G \in V$, a unique fixed point $\bar{s} = \bar{s}(G) \in \mathcal{S}^{0,1}(L)$ of $\mathcal{A}(\cdot,G)$, which depends continuously on $G \in V$ and satisfies $\bar{s}(F) = 0$. By formally differentiating relative to a the equation $\bar{s}(a)(\theta) = \mathcal{A}(\bar{s},G)(a,\theta)$, and using the definition of differentiability, we can prove the existence of a function which turns out to be continuous and equal to the derivative of $\bar{s}$ relative to a. This establishes that $\bar{s}$ is C^1 in a.

Define $g: \mathbb{R}^n \to \mathbb{R}^n$ by $g(b) = G[\xi(\cdot;b,f) + \bar{s}(G)(b)(\cdot)]$ and consider the set $S(G) = \{\varphi \in C^0: \varphi(\theta) = \xi(\theta;a,g), a \in \mathbb{R}\}$. Clearly, $S(G)$ is diffeomorphic to $\mathbb{R}^n$, depends continuously on G and the flow of (7.1) on $S(G)$ is given by the ODE $\dot{x}(t) = g(x(t))$. Consequently, in order to finish the proof, we need to show $B(G) = S(G)$.

Let

$$y(\theta) = \xi(\theta;a,f) + \bar{s}(G)(a)(\theta), \quad \theta \in [-r,0].$$

We have

$$y(\theta) = \xi(\theta;a,f) + \int_0^\theta H(\theta-\tau,y(\tau))[g(y(\tau)) - f(y(\tau))]d\tau$$

which is the nonlinear variation of constants formula for $\dot{x}(t) = f[x(t)] + [g(x(t)) - f(x(t))]$, and, therefore, we have

$$\xi(\theta;a,g) = \xi(\theta;a,f) + \bar{s}(G)(a)(\theta), \quad \theta \in [-r,0].$$

Also, using the last identity and the definition of g,

$$\begin{aligned}\dot{\xi}(t;a,g) = g(\xi(t;a,g)) &= G[\xi(\cdot;\xi(t;a,g),g) \\ &= G(\xi(t+\cdot;a,g)).\end{aligned}$$

Therefore, $\xi(t;a,g)$ is a global solution of (7.1) and $S(G)$ is invariant under (7.1), proving that $S(G) \subset B(G)$.

The rest of the proof is similar to the argument used for the analogous situation in Theorem 7.5. We begin with the proof that $S(G)$ is uniformly asymptotically stable under (7.1), by showing that there exist $\sigma, \beta > 0$ such that

$$\inf_{\psi \in S(G)} |x_t(\varphi,G) - \psi| \leq \sigma e^{-\beta t}, \quad t \geq r, \quad \varphi \in C^0.$$

Using this and the fact that $B(G)$ is the set of points in the global orbits of (7.1), we get $B(G) \subset S(G)$, and consequently, $B(G) = S(G)$.

Remark. Under certain conditions, the preceding proof can be generalized to situations where the unperturbed RFDE is not given by an ordinary differential equation, but there exists a submanifold S of the phase space where the flow is given by a C^1 ODE in $\mathbb{R}^n$, in the sense that there exists a C^1 function $h: \mathbb{R}^n \to \mathbb{R}^n$ such that the ODE $\dot{x}(t) = h(x(t))$ has unique solutions for each arbitrary initial condition $x(0) = a \in \mathbb{R}^n$ and that its solutions coincide with the solutions of the unperturbed RFDE $\dot{x}(t) = F(x_t)$ which have initial data on S (such manifolds are necessarily diffeomorphic to $\mathbb{R}^n$).

8. Stability Relative to A(F) and Bifurcation

As for ordinary differential equations, the primary objective in the qualitative theory of RFDEs is to study the dependence of the flow $\Phi_t = \Phi_t^F$ on F. This implicitly requires the existence of a criterion for deciding when two RFDEs are equivalent. A study of the dependence of the flow on changes of the RFDE through the use of a notion of equivalence based on a comparison of all orbits is very difficult and is likely to give too small equivalence classes. The difficulty is associated with the infinite dimensionality of the phase space and the associated smoothing properties of the solution operator. In order to compare all orbits of two RFDEs one needs to take into account the changes in the range of the solution map Φ_t, for each fixed t, a not so easy task due to the difficulties associated with backward continuation of solutions. Therefore, it is reasonable to begin the study by considering a notion of equivalence which ignores some of the orbits of the RFDEs to be compared. We restrict ourselves to RFDEs defined by functions $F \in \mathscr{X}^1$.

As in ODE's, the equilibrium points and periodic orbits play a very important role in the qualitative theory. In showing that two ODE's are equivalent, a fundamental role is played by linearization of the flow near equilibrium points and linearization of the Poincaré map near a periodic orbit - the famous Hartman-Grobman theorem. What is the generalization of this result for RFDE's? To see some of the difficulties, we consider equilibrium points in some detail.

Suppose p_F is an equilibrium point of an RFDE(F). If p_F is hyperbolic as a solution of F, then an application of the Implicit Function

Theorem guarantees the existence of neighborhoods U of F in $\mathscr{X}^1$ and V of p_F in C^0 such that, for each $G \in U$, there is a unique equilibrium point p_G in V and it is hyperbolic. Furthermore, the local stable manifold $W^s_{loc}(p_F)$ and local unstable manifold $W^u_{loc}(p_F)$ of p_F are diffeomorphic to the corresponding ones for G. The fact that these sets are diffeomorphic does not necessarily imply that the flows are equivalent in the sense that all orbits of F near p_F can be mapped by a homeomorphism onto orbits of G near p_G. The smoothing property of the flow generally prevents such a homeomorphism from being constructed. This implies the Hartman-Grobman theorem will not be valid; that is, the flow cannot be linearized near p_F. On the other hand, the local unstable manifolds are finite dimensional and, consequently, the restriction of the flows to them can also be described by ordinary differential equations. It follows that the flows $\Phi^F_t/W^u_{loc}(p_F)$ and $\Phi^G_t/W^u_{loc}(p_G)$ are diffeomorphisms, and therefore one can find a homeomorphism $h\colon W^u_{loc}(p_F) \to W^u_{loc}(p_G)$ which preserves orbits. The proof that such an h exists follows along the same lines as the proof of the classical Hartman-Grobman theorem making use of the analytic representation of the unstable manifolds $W^u_{loc}(p_F)$ and $W^u_{loc}(p_G)$ in terms of a coordinate system on the linearized unstable manifolds. The details of this proof were communicated to the authors by Jurgen Quandt.

It is also possible to define the global unstable set $W^u(p_F)$ of p_F by taking the union of the orbits through $W^u_{loc}(p_F)$. However, if $\Phi_t(F)$ is not one-to-one, then the manifold structure may be destroyed. On the other hand, if $\Phi_t(F)$ is one-to-one and $D\Phi_t(F)$ is one-to-one, then $W^u(p_F)$ is a finite dimensional immersed submanifold of $C^0(I,M)$.

The attractor set $A(F)$ contains all ω-limit and α-limit points of bounded orbits of F, as well as the equilibrium points, the periodic orbits and the bounded unstable manifolds of both. As a matter of fact, $A(F)$ consists of all the points of orbits of solutions that have a backward continuation and, thus, it is reasonable to begin the qualitative theory by agreeing to make the definitions of equivalence relative to the attractor set $A(F)$.

If $A(F)$ is compact and Φ_t is one-to-one on $A(F)$, then Φ_t is a group of $A(F)$. This implies that the solution operator does not smooth on $A(F)$. Therefore, one can attempt to modify several of the important ideas and concepts from ordinary differential equations so they are meaningful for RFDE's. These remarks suggest the following definition.

Definition 8.1. Two RFDEs F and G defined on manifolds are said to be equivalent, $F \sim G$, if there is a homeomorphism $h: A(F) \to A(G)$ which preserves orbits and sense of direction in time. An RFDE(F) defined on a manifold is said to be A-stable if there is a neighborhood V of F such that $G \sim F$ if $G \in V$.

As mentioned in Section 3, every ordinary differential equation on a manifold M can be considered as an RFDE on M with phase space $C^0(I,M)$. In particular, if X is a vector field on M and $\rho: C^0(I,M) \to M$ is the evaluation map $\rho(\varphi) = \varphi(0)$, the function $F = X \circ \rho$ is an RFDE on M. For each point $p \in M$ there is a solution of the ordinary differential equation defined by the vector field X which passes through p at $t = 0$. The map $\Sigma_X: M \to C^0(I,M)$ such that $\Sigma_X(p)$ is the restriction to $I = [-r,0]$ of the solution of X through p at $t = 0$, is a cross-section with respect to ρ and the attractor set of F is a manifold diffeomorphic to M and given by $A(F) = \Sigma_X(M)$. Clearly, the qualitative behavior of the flow of F on $A(F)$ is in direct correspondence with the qualitative behavior of the flow of the ordinary differential equation defined by X on M. It follows that all the bifurcations that occur for ordinary differential equations also occur for RFDEs. In this sense, the definition of A-stability given above is a generalization of the usual definition for ordinary differential equations.

Does an analogue of the Hartman-Grobman theorem hold if we restrict the flow Φ_t to the attractor set $A(F)$, as suggested above? In the infinite dimensional case, this always will lead to difficulties and they occur even in some finite dimensional problems.

It is natural to attempt to formulate the Hartman-Grobman theorem in the following way. For an hyperbolic equilibrium point p_F of the RFDE(F) suppose $\dim W^u_{loc}(p_F) = q$ and suppose there is a neighborhood U of p_F such that $\dim(A(F) \cap U) = r = q+s$, $s \geq 0$. Choose the smallest nonnegative integer $\bar{s} \geq s$ such that there is a $\mu > 0$ such that the number of eigenvalues of the linear variational equation about p_F with real parts in $[-\mu,0]$ is exactly $\bar{s}$. Now try to show that, generically in F, the exponential rate of attraction of any orbit in $A(F)$ towards the equilibrium point p_F is of order $e^{\beta t}$, $\beta \in [-\mu,0)$. If this can be done, then linearize the flow in $C^0(I,M)$ in the direction of the subspace of dimension $q + \bar{s}$ corresponding to the eigenfunctions of the eigenvalues with real parts $\geq -\mu$. Now identify the orbits in $A(F) \cap U$ with the linearized flow.

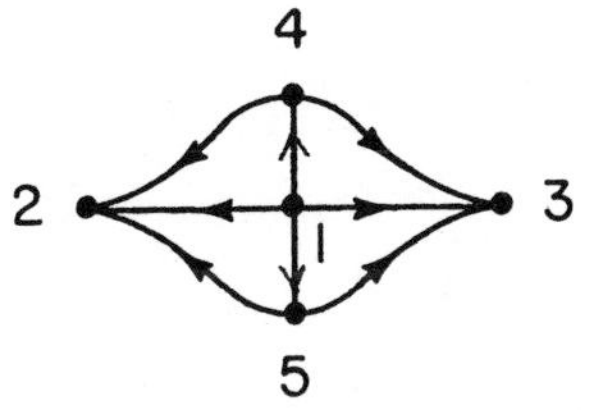

FIGURE 8.1

One must say "generic in F" in the above formulation for the following reason. Suppose A(F) as shown in Figure 8.1, with $\dim[W^s_{loc}(p^2_F) \cap A(F)] = 2$. Then the complete description of the A(F) near p^2_F must be determined by at least two eigenvalues of the linearized equation. Since nothing is known about the detailed structure of A(F), one would expect, generically in F, that the exponential behavior at A(F) could be determined by two eigenvalues with largest negative real part.

It is precisely the fact that one must say generically in F to formulate a reasonable Hartman-Grobman theorem that seems to make it impossible to state one. When one says generically in F, the global structure of the flow on A(F) begins to play a role. In fact, one can construct an A(F) as in Figure 8.2 where the points P_4 is semistable and there is an open set U of G with $F \in \partial U$ such that A(G) consists only of the points P_1, P_2, P_3 with the complete flow on A(G) is given as in Figure 8.3. The dimension of the local stable manifold of P_1 and P_4 in A(G) is only one whereas in A(F), it was two. This implies that a generalization of the Hartman-Grobman (if it exists) will require some new ideas.

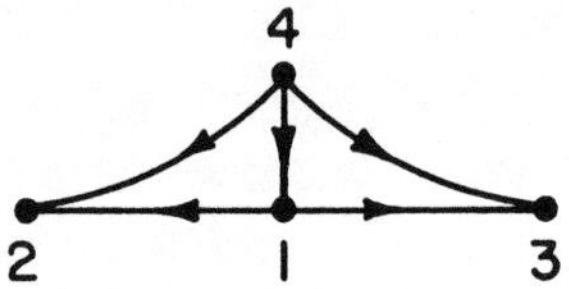

FIGURE 8.2

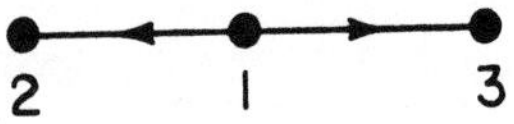

FIGURE 8.3

These examples also seem to indicate that the comparison of the flows of $A(F)$ and $A(G)$ near an equilibrium point will involve global properties of the flow.

We end this section with some examples from FDE's and elementary PDE's illustrating how the set $A(F)$ may vary with F and, in particular, how elementary bifurcations (non A-stable F) influence the behavior of $A(F)$. These special examples are chosen because they are nontrivial and yet it is still possible to discuss $A(F)$. Also, they illustrate the importance that the form of the equations play in the generic theory.

Let $b: [-1,0]$ be a C^2-function such that $b(\theta) > 0$, $\theta \in (-1,0]$, $b(-1) = 0$. Let $g: \mathbb{R} \to \mathbb{R}$ be a C^1-function and consider the equation

$$\dot{x}(t) = -\int_{-1}^{0} b(\theta)g(x(t+\theta))d\theta. \tag{8.1}$$

Proposition 8.2. *If* $b(0) \neq 0$ *and* $b'(\theta) \geq 0$, $b''(\theta) \leq 0$ *for* $\theta \in [-1,0]$, $G(x) = \int_0^x g \to \infty$ *as* $|x| \to \infty$, *then every solution of* 8.1 *is bounded and*

(i) *if there is a* $\theta_0 \in [-1,0]$ *such that* $b''(\theta_0) < 0$, *then every solution approaches a constant function as* $t \to \infty$, *the constant being a zero of* g.

(ii) _if_ $b''(\theta) = 0$ _for all_ θ _(that is,_ b _is linear) then, for any_ $\varphi \in C^0$, _there is either an equilibrium point or a one-periodic solution_ $p = p(\varphi)$ _of the ordinary differential equation_

$$\ddot{y} + a(0)g(y) = 0$$

such that the ω_-limit set of the orbit through_ φ _is_ $\{p_t, t \in \mathbb{R}\}$, _where_ $p_t \in C^0$, $p_t(\theta) = p(t+\theta)$, $-1 \leq \theta \leq 0$.

Let us first consider the case of a Hopf bifurcation. Suppose $xg(x) > 0$ for $x \neq 0$, $g'(0) = 1$. The linear variational equation of (8.1) for the zero solution is

$$\dot{x}(t) = -\int_{-1}^{0} b(\theta)x(t+\theta)d\theta$$

for which the characteristic equation is

$$\lambda + \int_{-1}^{0} b(\theta)e^{\lambda\theta}d\theta = 0.$$

If $b = b_0(\theta) = 4\pi^2(\theta+1)$, then this equation has two eigenvalues on the imaginary axis and the remaining ones have negative real parts. Furthermore, the set of b near b_0 for which this equation has two pure imaginary roots is a submanifold Γ of codimension one in the space of all b with the C^2-topology. One would expect that generically in g, there would be a Hopf bifurcation as one crosses Γ transversally. However, there is no generic Hopf bifurcation for any g.

This remark illustrates the difficulties that occur in the generic theory when the form of the differential equation is restricted.

Now let us consider the case where $b(\theta) > 0$, $-1 < \theta \leq 0$, $b(-1) = 0$, $b'(\theta) \geq 0$, $b''(\theta) \leq 0$ for $\theta \in [-1,0]$ and there exists a $\theta_0 \in [-1,0]$ such that $b''(\theta_0) < 0$. In this case, the ω-limit set of every solution of (8.1) is a zero of g and, also, the α-limit set of any nonconstant bounded solution of (8.1) is an unstable zero of g. If a is a zero of g then a is hyperbolic if and only if $g'(a) \neq 0$, uniformly asymptotically stable if $g'(a) > 0$ and unstable if $g'(a) < 0$. Furthermore, the unstable manifold $W^u(a)$ of a is one dimensional if a is unstable.

If the set of zeros of g is bounded, then there is a bounded set B such that every solution eventually enters B, that is, (8.1) is point dissipative. It follows (see Theorem 5.3) that there is a maximal compact invariant set $A_{b,g}$ for (8.1) which is uniformly asymptotically stable and attracts bounded sets of C.

From the fact that the α-limit set of any nonconstant bounded solution is an unstable zero of g, it follows that $A_{b,g} = \cup\{W^u(a): g(a) = 0\}$ and $A_{b,g}$ is one dimensional. To discuss the structure of the set $A_{b,g}$ for a fixed b and a certain class of g, let G_k be the class of all C^1-functions g satisfying the following conditions:

1) $\int_0^x g(s)ds \to \infty$ as $|x| \to \infty$.

2) g has exactly $2k+1$ zeros $a_1 < a_2 < \dots < a_{2k+1}$ all of which are simple.

Let the topology on G_k be that generated by the seminorms $||g||_M = \sup_{x\in M}(|g(x)| + |g'(x)|)$, where M is a compact set in $\mathbb{R}$. For any

$g \in G_k$, all zeros of g are hyperbolic and the zeros a_{2j}, $j = 1,2,\ldots,k$, are saddle points with unstable manifolds $W^u(a_{2j})$ one dimensional. Thus, for each a_{2j}, there are exactly two distinct orbits defined for $t \in (-\infty,\infty)$ whose α-limit sets are a_{2j}. We call these orbits <u>emanating</u> from a_{2j}. Fix b as above. Let $g,\tilde{g} \in G_k$ have $a_1 < \ldots < a_{2k+1}$ and $\tilde{a}_1 < \ldots < \tilde{a}_{2k+1}$, resp., as their zeros. Call g and $\tilde{g}$ equivalent $(g \sim \tilde{g})$ if for all $i,j \in \{1,\ldots,2k+1\}$, there is an orbit $x(t)$ of (8.1) emanating from a_i and tending to a_j as $t \to \infty$ if and only if there is an orbit $x(t)$ of (8.1) emanating from $\tilde{a}_i$ and tending to $\tilde{a}_j$ as $t \to \infty$. This clearly defines an equivalence relation on G_k. We say $g \in G_k$ is <u>~-stable</u> if the equivalence class of g is a neighborhood of g in G_k.

It is not difficult to show that g is ~stable if the ω-limit set of every orbit in $A_{b,g}$ which is not a point is a stable zero of g; that is, a point a_n, n odd, $1 \leq n \leq 2k+1$. Since $A_{b,g}$ is a global attractor and uniformly asymptotically stable, this is equivalent to saying that g is ~-stable if the ω-limit set of every orbit of (1.1)(b,g) defined and bounded on $(-\infty,\infty)$ is a stable zero of g. If it were known that the map $\Phi_{b,g}(t)$ is one-to-one on $A_{b,g}$, this latter statement would be equivalent to the following: there is a neighborhood V of g such that, for any $\tilde{g} \in V$, there is a homeomorphism of $A_{b,g}$ onto $A_{b,\tilde{g}}$ which preserves orbits and sense of direction in time; that is, g is <u>stable relative to</u> $A_{b,g}$. We have not been able to prove that $T_{b,g}(t)$ is one-to-one on $A_{b,g}$ and this is the reason for taking the weaker definition of equivalence. If g is analytic, then $\Phi_{b,g}(t)$ is one-to-one.

The ultimate objective would be to describe the equivalence classes in G_k. The cases $k = 0,1$ are trivial. Suppose $k = 2$; that is, each

$g \in G_2$ has five zeros $a_1 < a_2 < a_3 < a_4 < a_5$ with a_2, a_4 being saddle points, and a_1, a_3, a_5 being uniformly asymptotically stable. If a_j is an unstable equilibrium point with a_k, a_ℓ being the corresponding ω-limit sets of the orbits emanating from a_j, we designate this by $j[k,\ell]$. The structure of the flow on $A_{b,g}$ and the equivalence classes in G_2 are then determined by a pair $\{2[i,j], 4[k,\ell]\}$ expressing the fact that the unstable manifold through a_2 has ω-limit set $\{a_i, a_j\}$ and the one through a_4 has ω-limit set $\{a_k, a_\ell\}$.

The result states there are exactly five equivalence classes in G_2; namely $\{2[1,3],4[3,5]\}$, $\{2[1,4],4[3,5]\}$, $\{2[1,5],4[3,5]\}$, $\{2[1,3],4[2,5]\}$, $\{2[1,3],4[1,5]\}$. The only class that preserves the natural order of the reals on $A_{b,g}$ is $\{2[1,3],4[3,5]\}$. The first, third and fifth case are $\sim$-stable. The second and fourth cases have a connection between the saddle points a_2 and a_4. It seems plausible that these cases are not $\sim$-stable, but no proof is available.

The fact that five equivalence classes can occur indicates clearly the importance of studying the structure of the flow on $A_{b,g}$ rather than merely asserting that every solution of (8.1) approaches a zero of g.

If $g(x) = g(x + 2\pi)$, then the above equation (8.1) can be interpreted as an RFDE on a circle S^1. The separate situations for $A_{b,g}$ can then be depicted as in the figure below if we identify two of the zeros a_1 and a_5 of g.

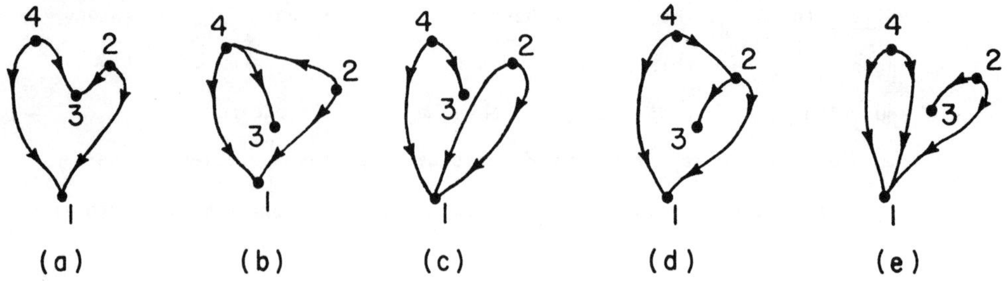

As another example, let us consider a simple parabolic equation. Consider the scalar equation

$$\begin{aligned} u_t &= u_{xx} + \lambda f(u), \quad 0 < x < \pi, \\ u &= 0 \quad \text{at} \quad x = 0, \pi \end{aligned} \tag{8.2}$$

with $\lambda > 0$ being a real parameter and $f(u)$ being a given non-linear function of u. If

$$V(\varphi) = \int_0^\pi [\varphi_x^2 - \lambda F(\varphi)]dx, \quad F(u) = \int_0^u f, \tag{8.3}$$

and $u(t,x)$ is a solution of (2.1), then

$$\frac{d}{dt}V(u(t,x)) = -\int_0^\pi u_t^2 dx \leq 0. \tag{8.4}$$

Theorem 8.4. *If*

$$F(u) \to -\infty \quad \text{as} \quad u \to \pm\infty \tag{8.5}$$

then Eq. (8.2) *generates a* C_0-*semigroup* $T_\lambda(t)$, $t \geq 0$, *on* $X = H_0^1(0,\pi)$, *each orbit is bounded and has* ω-*limit set as an equilibrium point*. *There is a maximal compact invariant set* A_λ *for* $T_\lambda(t)$ *which has the stability properties mentioned in Theorem* 5.3. *Finally, if* $\varphi \in A_\lambda$, *then the* α-*limit set of* φ *is an equilibrium point*.

The equilibrium points of (8.2) are the solutions of the equation

$$\begin{aligned} u_{xx} + \lambda f(u) &= 0, \quad 0 < x < \pi \\ u &= 0 \quad \text{at} \quad x = 0, \pi \end{aligned} \tag{8.6}$$

Equation (2.1) generates a C_0-semigroup and the ω- and α-limit sets must be a single equilibrium point. Relation (8.5) implies the set of equilibrium points is bounded. Since every orbit approaches an equilibrium point, one obtains point dissipative.

An equilibrium point, u_0 is hyperbolic if no eigenvalue of the operator $\partial^2/\partial x^2 + \lambda f'(u_0)$ on X is zero and it is called stable (hyperbolic) if all eigenvalues are negative. The unstable manifold $W^u(u_0)$ is the set of $\varphi \in X$ such that $T_\lambda(t)\varphi$ is defined for $t \leq 0$ and $\to u_0$ as $t \to -\infty$. The stable manifold $W^s(u_0)$ is the set of $\varphi \in X$ such that $T_\lambda(t)\varphi \to u_0$ as $t \to \infty$. The set $W^u(u_0)$ is an imbedded submanifold of X of finite dimension m (m being the number of positive eigenvalues of the above operator). The set $W^s(u_0)$ is an imbedded submanifold of codimension m. These manifolds are tangent at u_0 to the stable and unstable manifolds of the linear operator $\partial^2/\partial x^2 + f'(u_0)$ on X.

The following remark is a simple but important consequence of Theorem 8.4.

Corollary 8.5. *If* (8.5) *is satisfied and there are only a finite number of hyperbolic equilibrium points* $\varphi_1, \varphi_2, \ldots, \varphi_k$ *of* (8.2) *with each being hyperbolic, then*

$$A_\lambda = \bigcup_{j=1}^{k} W^u(\varphi_j).$$

Corollary 8.5 states that A_λ is the union of a finite number of finite dimensional manifolds. The complete dynamics on A_λ will only be known when we know the specific way in which the equilibrium points are connected to each other by orbits.

It seems to be difficult to discuss the complete flow on A_λ in the general case. Therefore, let us consider the special case of equation (8.2) where

$$f(0) = 0, \quad f'(0) = 1$$
$$\limsup f(u)/u \leq 0, \quad uf''(u) < 0 \quad \text{if} \quad u \neq 0. \tag{8.7}$$

Theorem 8.6. *If f satisfies (8.7) and $\lambda \in (n^2,(n+1)^2)$, n an integer, then there are exactly $2n+1$ equilibrium points $\alpha_\infty = 0$, α_j^+, α_j^-, $j = 0,1,\ldots, n-1$, where α_j^+,α_j^- have j zeros in $(0,\pi)$, $\dim W^u(\alpha_j^\pm) = j$, $0 \leq j \leq n-1$, $\dim W^u(\alpha_\infty) = n$ and*

$$A_\lambda = (\bigcup_j W^u(\alpha_j^\pm)) \cup W^u(\alpha_\infty).$$

For $n^2 < \lambda \leq (n+1)^2$, $n = 0,1,2,3$, the attractor A_λ has the form shown in the accompanying Figure 8.5.

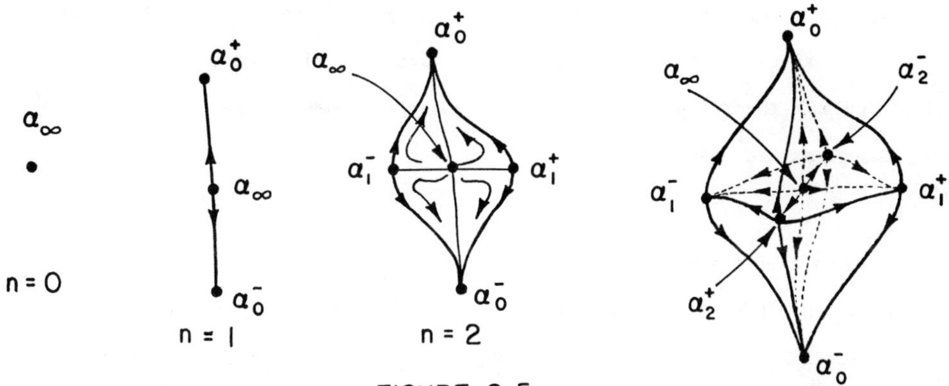

FIGURE 8.5

Hale and Nascimento in an unpublished manuscript have shown that for $n^2 < \lambda \leq (n+1)^2$ and arbitrary n, there exist orbits connecting α_∞ to $\alpha_j^\pm$ for all j and orbits from $\alpha_j^\pm$ to $\alpha_k^\pm$ for all $j > k$.

Another interesting example comes from a model for the transverse motion of an elastic beam with ends fixed in space which is given by the nonlinear equation

$$u_{tt} + \alpha u_{xxxx} - [\lambda + k\int_0^\ell u_s^2(s,t)ds]u_{xx} + \delta u_t = 0 \tag{8.8}$$

where α,β,δ and λ are positive constants and the boundary conditions are stated for hinged or clamped ends. In each case, the equation defines a flow in a suitable Banach space, with a Liapunov function (the energy) nonincreasing along solutions. Taking f as the time-one map, the number of fixed points and the dimension of $A(f)$ depend on λ (which is proportional to the axial load) (see Section 10).

It can be proved that the set $A(F)$ corresponding to the semigroup $T_F(t)$ generated by the Navier-Stokes equation F in a two dimensional domain is a compact set and $f = T_F(1)$ is a compact map. The dimension of $A(F)$ may increase as the Reynolds number Re increases and it seems that the dynamical system $F = F(Re)$ is in fact the object of investigation in turbulence theory dealing with flows at large values of Re. How does $A(F)$ change as $Re \to +\infty$?

The examples just mentioned are a good illustration of Morse-Smale systems in infinite dimensions. To make this concept precise, we need some notation: Let $\{T_f(t), t \geq 0\}$ be an abstract dynamical system as defined in Section 1. For any hyperbolic equilibrium point x of this system, one can define in the usual way the local stable manifold $W^s_{loc}(x)$ and the local unstable manifold $W^u_{loc}(x)$. The manifold $W^u_{loc}(x)$ has finite dimension. For any hyperbolic periodic orbit γ, one can also define the local stable manifold $W^s_{loc}(\gamma)$ and local unstable manifold $W^u_{loc}(\gamma)$ of γ, with the latter being finite dimensional. One can define the global unstable sets $W^u(x)$ and $W^u(\gamma)$ by taking the union of the orbits through points in $W^u_{loc}(x)$, $W^u_{loc}(\gamma)$, respectively. In the general case, these sets will not be manifolds. To be certain that $W^u(x)$, $W^u(\gamma)$ are immersed submanifolds, we suppose that $DT_f(t)$ is one-to-one on the tangent space of X at points of the attractor $A(f)$, for all t.

We can now make the following definition:

Definition 8.7. The dynamical system $\{T_f(t), t \geq 0\}$, is said to be Morse-Smale if

(i) $DT_f(t)$ is one-to-one on the tangent space of X at points of $A(f)$;

(ii) the nonwandering set $\Omega(f)$ is the union of a finite number of equilibrium points and periodic orbits, all hyperbolic;

(iii) the local stable and global unstable manifolds of all equilibrium points and periodic orbits intersect transversally.

The situations depicted in Figure 8.4 a), c), e) for Equation (8.1) are Morse-Smale systems as well as the ones shown in Figure 8.5 for Equation (8.2) and $\lambda \in (n^2,(n+1)^2)$, $n = 0,1,2,3$. Other examples are given in Section 9.

9. Compactification at Infinity

The behavior at infinity of solutions of ordinary differential equations in the plane was studied by Poincaré by compactification of the Euclidean plane into the unit two-dimensional sphere S^2. The same idea of compactification can be applied to RFDEs. In order to illustrate this, we present here a study on equations obtained by compactification of linear delay equations $\dot{x}(t) = Ax(t-1)$ in $\mathbb{R}^2$ and in $\mathbb{R}$ (compactified to the sphere S^2 and the circle S^1, respectively).

Consider, in the plane $\mathbb{R}^2$, the linear system of delay equations

$$\dot{x}(t) = Ax(t-1) \tag{9.1}$$

where A is a 2×2 real nonsingular matrix, and let $I = [-1,0]$. In order to describe the Poincaré compactification into the sphere $S^2 = \{y \in \mathbb{R}^3 : \sum_{i=1}^{3} y_i^2 = 1\}$, let us identify $\mathbb{R}^2$ with the plane $T_N S^2 = \{y \in \mathbb{R}^3 : y_3 = 1\}$, where $N = (0,0,1)$ is called the north pole of S^2. The Poincaré compactification is obtained by the maps that assign to each point $(x_1,x_2,1) \in T_N S^2$ the points of intersection of S^2 with the straight line passing through $(x_1,x_2,1)$ and the origin, i.e., it is given by two maps $\pi_i : T_N S^2 \to S^2$, $i = 1,2$, such that

$$\pi_i(x_1,x_2,1) = \frac{(-1)^i (x_1,x_2,1)}{(1+x_1^2+x_2^2)^{1/2}} .$$

By Poincaré compactification, equation (9.1) is transformed to an equation on S^2. The points at infinity in $\mathbb{R}^2$ are mapped onto the equator, i.e., $S^1 = \{(y_1,y_2,y_3) \in \mathbb{R}^3 : y_3 = 0\}$. It is reasonable to start studying the behavior at infinity by compactifying in such a way that the equator be invariant and the equation on S^2 be analytic. This can be accomplished by multiplying (9.1) by the factor

$$\frac{y_3(t-1)}{y_3(t)} = \left(\frac{1 + x_1^2(t) + x_2^2(t)}{1 + x_1^2(t-1) + x_2^2(t-1)}\right)^{1/2}.$$

The delay equation in S^2 obtained in this way is denoted by $\pi(A)$ and it is given by the restriction to S^2 of the following system on $\mathbb{R}^3$

$$\begin{pmatrix} \dot{y}_1(t) \\ \dot{y}_2(t) \\ \dot{y}_3(t) \end{pmatrix} = \begin{pmatrix} 1-y_1^2(t) & -y_1(t)y_2(t) \\ -y_1(t)y_2(t) & 1-y_2^2(t) \\ -y_1(t)y_3(t) & -y_2(t)y_3(t) \end{pmatrix} A \begin{pmatrix} y_1(t-1) \\ \\ y_2(t-1) \end{pmatrix} \tag{9.2}$$

The behavior at infinity in $\mathbb{R}^2$ is described by the restriction of (9.2) to the equator S^1, which can be written in polar coordinates for the plane $y_3 = 0$ as

$$\dot{\theta}(t) = (-\sin\theta(t), \cos\theta(t))\, A \begin{pmatrix} \cos\theta(t-1) \\ \sin\theta(t-1) \end{pmatrix} \tag{9.3}$$

If $A = (a_{ij})_{i,j=1}^2$, then the initial points of system (9.2) on S^2 are $N = (0,0,1)$, $S = (0,0,-1)$ and the points on the equator S^1 which correspond to solutions of

$$(a_{22}-a_{11})\sin\theta\cos\theta + a_{21}\cos^2\theta - a_{12}\sin^2\theta = 0.$$

We first give a generic result for $\pi(A)$.

Theorem 9.1. The set $\mathscr{A}$ of all 2×2 real nonsingular matrices A for which $\pi(A)$ on S^2 has all critical points hyperbolic is open and dense

<u>in the set</u> M(2) <u>of all real</u> 2×2 <u>matrices. Furthermore, if</u> $A \in \mathscr{A}$ <u>then it is equivalent under a similarity transformation to one of the following types of matrices:</u>

$$\text{(I)} \quad A = \begin{pmatrix} a_1 & 0 \\ 0 & a_2 \end{pmatrix} \quad \text{with } a_1 \neq a_2$$

$$\text{(II)} \quad A = \begin{pmatrix} \alpha & -\beta \\ \beta & \alpha \end{pmatrix}, \quad \beta > 0.$$

<u>Proof</u>: One first observes that the critical points in the equator are not hyperbolic if the eigenvalues of A are not distinct. This immediately implies that $\mathscr{A}$ contains either matrices of types I or II. One then shows that the set of all real nonsingular 2×2 matrices A with distinct eigenvalues is open and dense in M(2).

For matrices of type (I), the critical points are N,S and four points in the equator given in terms of the polar angle by $\theta = 0, \pi/2, 3\pi/2$ and π. The hyperbolicity of N and S is equivalent to $-a_1, -a_2 \neq (\pi/2 + 2n\pi)$, $n = 0, \pm 1, \pm 2$. The linear variational equation at the points on the equator given by $\theta = 0, \pi$ can be expressed in spherical coordinates defined by $y_1 = \cos\psi \cos\varphi$, $y_2 = \cos\psi \sin\varphi$, $y_3 = \sin\psi$, as

$$\dot{\psi}(t) = -a_1\psi(t)$$
$$\dot{\varphi}(t) = -a_1\varphi(t) + a_2\varphi(t-1).$$

The only possibility for characteristic values of these equations to belong to the imaginary axis is to have $|a_1| < |a_2|$ and then the characteristic values $\lambda = iy$ must satisfy $\cos y = a_1/a_2$ and $y^2 = a_2^2 - a_1^2$. Perturbing a_1 and a_2 with a_1/a_2 = constant, we obtain hyperbolicity. The points in the equator given by $\theta = \pi/2,\ 3\pi/2$ are treated in a similar way.

For matrices of type (II), there are no critical points in the equator and the characteristic values of the linear variational equation at N and S in the imaginary axis, $\lambda = iy$, must satisfy $y^2 = \alpha^2 + \beta^2$ and $\tan y = \pm\alpha/\beta$. Perturbing α and β while maintaining α/β = constant we obtain hyperbolicity.

In the case (II) of Theorem 9.1, by the use of spherical coordinates $y_1 = \cos\psi\cos\varphi$, $y_2 = \cos\psi\sin\varphi$, $y_3 = \sin\psi$, the equation $\pi(A)$ on S^2 can be written

$$\begin{aligned}\dot\psi(t) &= -(\alpha^2+\beta^2)^{1/2}\sin\psi(t)\cos\psi(t-1)[\cos\varphi_0-\varphi(t)+\varphi(t-1)]\\ \dot\varphi(t) &= (\alpha^2+\beta^2)^{1/2}\frac{\cos\psi(t-1)}{\cos\psi(t)}\sin[\varphi_0-\varphi(t)+\varphi(t-1)]\end{aligned} \tag{9.4}$$

where $0 < \varphi_0 < \pi$ satisfies

$$\cos\varphi_0 = (\alpha^2+\beta^2)^{-1/2}\alpha$$
$$\sin\varphi_0 = (\alpha^2+\beta^2)^{-1/2}\beta.$$

In the equator of S^2, we have

$$\dot\varphi(t) = (\alpha^2+\beta^2)^{1/2}\sin[\varphi_0 - \varphi(t) + \varphi(t-1)]. \tag{9.5}$$

Theorem 9.2. In the case (II) of Theorem 9.1, any periodic orbit of equation $\pi(A)$ in the equator of S^2 is given by a periodic solution of constant velocity. If $M = (\alpha^2+\beta^2)^{1/2} < 1$, then the set of all global solutions in the equator consists of exactly one asymptotically stable hyperbolic periodic solution. There exists a sequence $M_0 < M_1 < M_2 < \ldots,\ M_n \to \infty$, such that for $M_i < M < M_{i+1}$ there exist exactly $2i+1$ periodic orbits in the equator, their velocities are distinct with the highest velocity increasing to ∞ as M increases, and they are hyperbolic and alternatively asymptotically stable or unstable under the ordering of magnitude of these velocities. If $M = M_i$, $i > 1$, then there exist exactly $2i$ periodic orbits in the equator and all of them, except the one with highest speed are hyperbolic and alternately asymptotically stable or unstable.

Proof: Let $\varphi(t)$ be a T-periodic solution of (9.5). Then $u(t) = \varphi(t) - \varphi(t-1)$ satisfies $u(t) = u(t+T)$, and

$$\dot{u}(t) = -M[\sin(\varphi_0 - u(t)) - \sin(\varphi_0 - u(t-1))] \tag{9.6}$$

with $M = (\alpha^2+\beta^2)^{1/2}$. It can be shown that any solution $x(t)$ of $\dot{x}(t) = g(x(t)) - g(x(t-1))$ converges to a limit (finite or infinite) as $t \to \infty$ provided g is continuously differentiable. Besides, we can write

$$x(t) = x(0) - \int_{-1}^{0} g(x(\theta))d\theta + \int_{t-1}^{t} g(x(\tau))d\tau.$$

Since (9.6) is an equation of this type with $g(x) = -M \sin(u_0 - x)$, we get that $u(t)$ is bounded and converges to a constant as $t \to \infty$. Since $u(t) = u(t+T)$, $\varphi(t) - \varphi(t-1) = u(t)$ is a constant function, and, from equation (9.5), we have $\dot{\varphi}(t)$ also constant, proving the first statement.

If $\varphi(t)$ is a global solution in the equator, then, from (9.5) with $u(t) = \varphi(t) - \varphi(t-1)$, we have

$$u(t) = M \int_{t-1}^{t} \sin[\varphi_0 - u(\tau)]d\tau.$$

Consider the Banach space $\mathscr{B}$ of all real continuous bounded functions with the sup norm, and let $\mathscr{T}: \mathscr{B} \to \mathscr{B}$ be the map transforming u into the function of t given by the right-hand side of the preceding equation. We have $||\mathscr{T}(u_1) - \mathscr{T}(u_2)|| \leq M||u_1 - u_2||$. Thus, if $M < 1$, $\mathscr{T}$ is a contraction map and therefore there exists a unique fixed point u_0 of $\mathscr{T}$ in $\mathscr{B}$. Any solution ω of

$$\sin(\omega-\varphi_0) = -\omega/M \tag{9.7}$$

is a fixed point of $\mathscr{T}$ and there exists always at least one solution of this equation. Hence the function $u(t)$ is constant and, therefore, φ is 1-periodic.

To study the hyperbolicity of the periodic orbits in the equator, which we know have constant velocities, we consider the linear variational equation of (9.4) around solutions $\psi(t) = 0$, $\varphi(t) = \omega t$. Clearly, ω must satisfy equation (9.7). It is then easy to prove by analysis of characteristic values that a periodic orbit in the equator with velocity ω is hyperbolic if and only if $\cos(\varphi_0-\omega) \neq 0$ and $M \cos(\varphi_0-\omega) \neq -1$. From the study of the characteristic equation, it also follows that all characteristic values have negative real parts if $M \cos(\varphi_0-\omega) > 0$, and, therefore, the corresponding periodic orbits of constant velocity ω are asymptotically stable.

Since equation (9.7) describes the velocities of periodic orbits in the equator, one has only to study the roots of this equation to conclude the rest of the statement (see Fig. 9.1).

Remark. It is easy to see that the unstable manifolds of the hyperbolic unstable periodic orbits have dimension two. Also, it can be shown that, for an open and dense set of matrices such that $M < 3\pi/2$, the Equation (9.5) is Morse-Smale (see Section 8 for the definition).

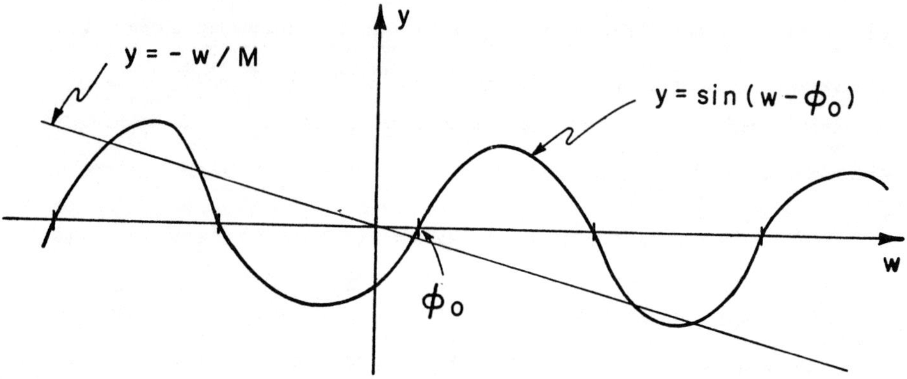

FIGURE 9.1

We now consider the scalar equation

$$\dot{x}(t) = -ku(t-1), \quad k \neq 0, \quad x(t) \in \mathbb{R}.$$

By Poincaré compactification, we can define an RFDE on the circle $S^1 = \{(y_1, y_2) \in \mathbb{R}^2 : y_1^2 + y_2^2 = 1\}$ using the projections defined by

$$\pi_i(x) = (-1)^i \frac{(x,1)}{(1+x^2)^{1/2}} .$$

In order to obtain an analytic equation on S^1 which leave the points corresponding to infinity invariant, we multiply (9.8) by the factor

$$\frac{y_2(t-1)}{y_2(t)} = \left(\frac{1+x^2(t)}{1+x^2(t-1)}\right)^{1/2}$$

before projecting into S^1. Introducing polar coordinates, we obtain

$$\dot{\theta}(t) = k \sin \theta(t) \cos \theta(t-1) \tag{9.9}$$

There exist four critical points corresponding to $\theta = 0, \pi/2, \pi, 3\pi/2$: $A = (1,0)$, $B = (0,1)$, $C = (-1,0)$, $D = (0,-1)$. The linear variational equation for the points corresponding to infinity, A and C, is $\dot{\theta} = k\theta$ and, therefore, the equation behaves like an ODE close to these points. At the poles B and D, the linear variational equation is precisely the original equation (9.8).

Theorem 9.3. There is a Hopf bifurcation for (9.9) near A and C for k near $(\pi/2 + 2n\pi)$, n integer, If $k > \pi/2$, then (9.9) has periodic solutions of period $T = 4$ satisfying the symmetry conditions $\theta(t) = \pi - \theta(t-2)$, $t \in \mathbb{R}$.

Proof: The first statement is a standard application of the Hopf bifurcation theorem.

For the second statement, assume $\theta(t)$ is a global solution of (9.9) and let $\varphi(t) = \theta(t) - \pi/2$, $\psi(t) = \varphi(t-1)$. Then $\dot{\psi}(t) = -k \cos \psi(t) \sin \varphi(t-2)$. If there exists a solution $\theta(t)$ such that $\theta(t) = -\theta(t-2)$, then $\varphi(t)$, $\psi(t)$ must satisfy

$$\begin{aligned} \dot{\varphi}(t) &= -k \cos \varphi(t) \sin \psi(t) \\ \dot{\psi}(t) &= k \sin \varphi(t) \cos \psi(t). \end{aligned} \tag{9.10}$$

Clearly, this system is Hamiltonian on the torus T^2 with energy function $E = k \cos \varphi \cos \psi$. The phase portrait of this system in the (φ,ψ)-plane has centers at the points $\varphi = m\pi$, $\psi = n\pi$, and saddles at the points $\varphi = \pi/2 + m\pi$, $\psi = \pi/2 + n\pi$, for m,n integers (see Fig. 9.2). The saddle connections are contained in vertical and horizontal lines in the (φ,ψ)-plane. When we go to the torus, we get four saddles and four centers. The limit period of the periodic orbits is $T_\ell = 2\pi/k$ as the orbits approach a center and is $+\infty$ as the orbits approach a saddle. Then there exist always periodic

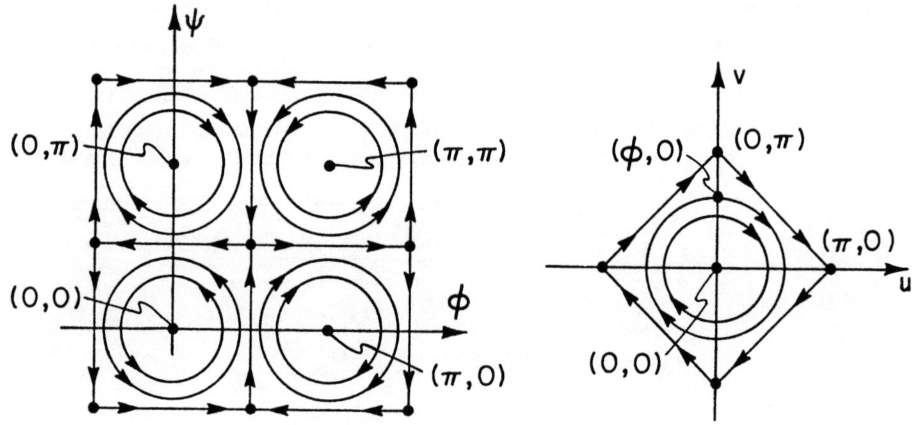

FIGURE 9.2

orbits with period greater than $2\pi/k$ and, since $k > \pi/2$, there exist periodic orbits with period four.

If we introduce new variables φ and ψ by the relations

$$u = \varphi+\psi \qquad \varphi = \frac{u+v}{2}$$
$$\text{or}$$
$$v = \varphi-\psi \qquad \psi = \frac{u-v}{2}$$

then system (9.10) becomes the Hamiltonian system

$$\begin{aligned} \dot{u} &= k \sin v = \frac{\partial E}{\partial v} \\ \dot{v} &= -k \sin u = -\frac{\partial E}{\partial u} \end{aligned} \tag{9.11}$$

where the energy is $E = -k(\cos u + \cos v)$.

We look for periodic solutions of (9.10) satisfying

$$\begin{array}{lcl} \psi(t) = \varphi(t-1) & & \varphi(t) = -\psi(t-1) \\ & \text{which imply} & \\ \varphi(t) = -\varphi(t+2) & & \psi(t) = -\psi(t+2) \end{array}$$

These conditions for a solution of (9.10) are equivalent to finding a periodic solution $(u(t),v(t))$ of (9.11) satisfying

$$\begin{array}{lcl} u(t) = v(t-1) & & u(t-1) = -v(t) \\ & \text{which imply} & \\ v(t+2) = -v(t) & & u(t) = -u(t+2). \end{array}$$

Now choose $k > \pi/2$ and, for simplicity, work in the square $|u| + |v| = \pi$ which contains four saddle connections of the (u,v) plane. There exists a $c > 0$ such that the solution defined by $v(1) = c$, $u(1) = 0$, $0 < c < \pi$, has period equal to 4.

Let $u(t)$, $v(t)$ be such a solution. Consider now the functions $\bar{u}(t) = -v(t)$ and $\bar{v}(t) = u(t)$ and verify that $(\bar{u}(t),\bar{v}(t))$ satisfy (9.11). But the solutions $(u(t),v(t))$ and $(\bar{u}(t)\bar{v}(t))$ have the same energy E since $E = -k(\cos u(t) + \cos v(t)) = -k(\cos \bar{u}(t) + \cos \bar{v}(t)) = -k(1 + \cos c)$. Thus, for a certain t^*, we have $\bar{u}(t^*) = 0$ and $\bar{v}(t^*) = c$. Both solutions define then the same periodic orbit and there exists $\alpha \in (0,4)$ such that

$$(u(t),v(t)) = (\bar{u}(t+\alpha),\bar{v}(t+\alpha)) = (-v(t+\alpha),u(t+\alpha)) \ \forall t \in \mathbb{R}.$$

But $u(t) = -v(t+\alpha) = -u(t+2\alpha) = v(t+3\alpha) = u(t+4\alpha)$ and, since $u(t)$ has period 4, we need to have $\alpha = 1$. Then $(u(t),v(t))$ satisfy the conditions required above since

$$u(t) = v(t+3) = v(t-1)$$

$$v(t+2) = u(t+3) = u(t-1) = -v(t).$$

The corresponding 4-periodic functions $\varphi(t), \psi(t)$ are such that $\theta(t)$ is a periodic solution of the equation (9.9) with period $T = 4$ and such that $\theta(t) = \pi - \theta(t-2)$.

10. Stability of Morse-Smale Maps

We will deal in this section with smooth maps $f: B \to E$, B being a Banach manifold imbedded in a Banach space E. The maps f belong to $C^r(B,E)$, the Banach space of all E-valued C^r-maps defined on B which are bounded together with their derivatives up to the order $r \geq 1$. Let $C^r(B,B)$ be the subspace of $C^r(B,E)$ of all maps leaving B invariant, that is, $f(B) \subset B$. Denote by $A(f)$ the set

$$A(f) = \{x \in B\text{: there exists a sequence } (x = x_1, x_2, \ldots) \in B,$$
$$\sup_j \|x_j\| < \infty \text{ and } f(x_j) = x_{j-1},\ j = 2,3,\ldots\}.$$

Special subspaces $KC^r(B,B)$ of $C^r(B,B)$ will be introduced satisfying the following compactness and reversibility properties: "any $f \in KC^r(B,B)$ is reversible, has $A(f)$ compact and given a neighborhood U of $A(f)$ in B, there exists a neighborhood $\mathscr{W}(f)$ of f in $KC^r(B,B)$ such that $A(g) \subset U$ for all $g \in \mathscr{W}(f)$" (reversibility for a C^1 map f means $f/A(f)$ and $df/A(df)$ are injective maps). The choice of the classes $KC^r(B,B)$ depends on the problems in view. In each case we need to assume appropriate hypothesis on the data in order to obtain the required compactness and reversibility properties for the elected $KC^r(B,B)$.

Global unstable manifolds of hyperbolic periodic orbits of a map $f \in KC^r(B,B)$ are introduced using the reversibility of f. The nonwandering set $\Omega(f)$ is the set of all $z \in A(f)$ such that given a neighborhood V of z in $A(f)$ and $n_0 \in N$, there exists $n > n_0$ such that $f^n(V) \cap V \neq \emptyset$. If $f \in KC^r(B,B)$, $\Omega(f)$ is compact and invariant and contains all $\omega(x)$ and $\alpha(x)$ of all $x \in A(f)$. Morse-Smale maps will be introduced (see Definition

10.14) and we denote by MS the set of Morse-Smale maps of $KC^r(B,B)$.

From the dynamics point of view, we will see that a Morse-Smale map f exhibits the simplest orbit structure, specially the "gradient like" ones, that is, the $f \in MS$ for which there exists a continuous Liapunov function $V: B \to R$ such that if $x \in B$ and $f(x) \neq x$, then $V(f(x)) < V(x)$. In this case $\Omega(f)$ is equal to $Fix(f)$, the set of all fixed points of f.

Important stability theorems for (and existence of) Morse-Smale diffeomorphisms defined on a compact manifold M are well known. They say that any Morse-Smale diffeomorphism f is stable. That is, there exists a neighborhood $\mathcal{V}(f)$ of f in $Diff^r(M)$, the set of all C^r-diffeomorphisms of M, $r \geq 1$, such that for each $g \in \mathcal{V}(f)$ corresponds a homeomorphism $h = h(g): M \to M$ and $h \cdot f = g \cdot h$ holds on M.

We say that $f \in KC^r(B,B)$ is A-stable if there exists a neighborhood $\mathcal{V}(f)$ of f in $KC^r(B,B)$ such that to each $g \in \mathcal{V}(f)$ corresponds a homeomorphism h, $h = h(g): A(f) \to A(g)$ and $h \cdot f = g \cdot h$ holds on $A(f)$. The main results of this section can be summarized as follows:

"The set MS is open in $KC^r(B,B)$ and any $f \in MS$ is A-stable".

Let $x = f(x)$ be a fixed point of a C^r-map, $f: B \to B$, $r \geq 1$. The fixed point is said to be hyperbolic if the spectrum $\sigma(df(x))$ of the derivative $df(x)$ is disjoint from the unit circle of the complex plane. Under the above hypothesis one can define local unstable and local stable C^r-manifolds denoted by $W^u_{loc}(x)$ and $W^s_{loc}(x)$, respectively.

<u>Proposition 10.1.</u> <u>Let</u> $x = f(x)$ <u>be a hyperbolic fixed point of a reversible</u> C^r <u>map</u> $f: B \to B$. <u>The set</u> $W^u(x) = \bigcup_{i \geq 0} f^i(W^u_{loc}(x))$ <u>is an injectively immersed</u> C^r<u>-submanifold of</u> B.

The manifold $W^u(x)$ is the (global) unstable manifold of the hyperbolic fixed point x. It is easy to see that $W^u(x)$ is invariant under f so $W^u(x) \subset A(f)$.

For instance, if the given C^r map $f: B \to B$ is compact, the derivative $df(x)$ at the hyperbolic fixed point x is a linear compact operator and $W^u(x)$ is finite dimensional; $W^s_{loc}(x)$ is finite codimension and positively invariant. The manifolds $W^s_{loc}(x)$ and $W^u_{loc}(x)$ are always transversal at the point x.

If $g = f^n$, $n \geq 1$, is a power of a bounded map $f: B \to B$, it is easy to see that $A(f) = A(g)$. If $f/A(f)$ is injective then $g/A(g)$ is also injective. If f is compact, g is compact and if f is reversible, g is reversible.

$x \in B$ is a periodic point of f if it is a fixed point of some iterate of f; the smallest integer $m > 0$ with $f^m(x) = x$ is the period of x. It is clear that the orbit $\mathcal{O}(x) = \{x, f(x), f^2(x), \ldots, f^{m-1}(x)\}$ of a periodic point x is a finite set with m points. $\mathrm{Fix}(f)$ and $\mathrm{Per}(f)$ will denote, respectively, the set of all fixed points and of all periodic points of f. We have, obviously, $\mathrm{Fix}(f) \subset \mathrm{Per}(f) \subset \Omega(f)$.

A periodic point x with period m is said to be a hyperbolic periodic point if $\mathcal{O}(x)$ is hyperbolic, that is, if all points $y \in \mathcal{O}(x)$ are hyperbolic fixed points of f^m. We can talk about $W^u_{loc}(y)$, $W^s_{loc}(y)$ for all $y \in \mathcal{O}(x)$. The unstable manifold of y is $W^u(y) = \bigcup_{i \geq 0} f^{mi}(W^u_{loc}(y))$.

Definition 10.2. A hyperbolic periodic point x of f is a source if $W^s_{loc}(x) \cap A(f) = \{x\}$; is a sink if $W^u_{loc}(x) = \{x\}$; otherwise x is a saddle.

Proposition 10.3. Let f be a smooth C^0-reversible map, that is, $f/A(f)$ is injective, and x be a hyperbolic periodic source (sink; saddle). Then $y \in \mathcal{O}(x)$ is also a source (sink; saddle).

Let x be a hyperbolic fixed point of a smooth map $f: B \to B$ and assume $\dim W^u_{loc}(x) < \infty$. If x is not a sink there exists an open disc B^u in $W^u_{loc}(x)$ such that $\mathrm{Cl}\, B^u \subset W^u_{loc}(x)$ and f^{-1}/B^u is a contraction. It follows that $f^{-1}(B^u) \subset B^u$. As usually, a fundamental domain for $W^u_{loc}(x)$ is the compact set $G^u(x) = \mathrm{Cl}\, B^u - f^{-1}(B^u)$. If $y \in W^u_{loc}(x) - \{x\}$, there exists an integer k such that $f^k(y) \in G^u(x)$. Any neighborhood $N^u(x)$ of $G^u(x)$ such that $N^u(x) \cap W^s_{loc}(x) = \Phi$ is called a fundamental neighborhood for $W^u_{loc}(x)$.

If the hyperbolic fixed point is not a source we will consider a neighborhood $V = B^s \times B^u$ of x, B^s being an open disc in $W^s_{loc}(x)$ such that f/B^s is a contraction and $\mathrm{Cl}\, B^s \subset W^s_{loc}(x)$.

We define the fundamental domain for $W^s_{loc}(x)$ as

$$G^s(x) = \mathrm{Cl}[B^s \cap A(f)] - f(B^s \cap A(f)).$$

If $A(f)$ is compact and f is C^0-reversible, then $f/A(f)$ is a homeomorphism and $G^s(x)$ is compact. It is clear that $x \notin G^s(x)$ so $W^u_{loc}(x) \cap G^s(x) = \Phi$ and there exists a neighborhood $N^s(x)$ of $G^s(x)$ which does not intersect $W^u_{loc}(x)$; $N^s(x)$ is called a fundamental neighborhood for $W^s_{loc}(x)$.

Remarks: 1) Any point of $W^s \cap A(f) = [\bigcup_{i\geq 0} f^{-i}(W^s_{loc}(x))] \cap A(f)$ reaches $B^s \cap A(f)$ after finitely many iterations of $f/A(f)$ or its inverse.

2) Given $y \in B^s \cap A(f) - \{x\}$, there exists an integer k such that $\tilde{f}^k(y) \in G^s(x)$, $\tilde{f}$ being the restriction of f to $A(f)$.

In fact, if $y \notin f(B^s \cap A(f))$ there is nothing to prove. If $y \in f(B^s \cap A(f))$ one considers the sequence $y = y^0, y^1, y^2, \ldots, f(y^i) = y^{i-1}$, $i \geq 1$, and there exists a first integer i_0 such that $y^{i_0} \notin B^s \cap A(f)$ (if $y^i \in B^s \cap A(f)$ for all $i \geq 1$ then $y \in W^u_{loc}(x) \cap B^s = \{x\}$). If $y^{i_0} \in \mathrm{Cl}[B^s \cap A(f)]$, $y^{i_0} \in G^s(x)$; if $y^{i_0} \notin \mathrm{Cl}[B^s \cap A(f)]$ then $y^{i_0-1} \in B^s \cap A(f) - f(B^s \cap A(f)) \subset G^s(x)$.

Given two submanifolds $i_1: W_1 \to B$ and $i_2: W_2 \to B$ one says that W_1 and W_2 are $\varepsilon - C^1$ close manifolds if there exists a diffeomorphism $\gamma: W_1 \to W_2$ such that $i_1: W_1 \to B$ and $i_2 \circ \gamma: W_1 \to B$ are $\varepsilon - C^1$ close maps.

Proposition 10.4. (local λ-lemma). Let x be a hyperbolic fixed point of a smooth map $f: B \to B$, $\dim W^u_{loc}(x) < \infty$, and B^u be an imbedded open disc in $W^u_{loc}(x)$, containing x. Let q be a point of $W^s_{loc}(x)$, $q \neq x$, and D^u be a disc centered at q, transversal to $W^s_{loc}(x)$, such that $\dim D^u = \dim W^u_{loc}(x)$. Then there is an open set V of B containing B^u such that given $\varepsilon > 0$ there exists $n_0 \in N$ such that if $n > n_0$ the connected component of $f^n(D^u) \cap V$ through $f^n(q)$ and the open disc B^u are $\varepsilon - C^1$ close manifolds.

It is interesting to remark that we do not need to assume compactness or reversibility for the smooth map f but the available proofs use, strongly, the finite dimensionality of $W^u_{loc}(x)$.

In the same hypothesis of the local λ-lemma, let x be a hyperbolic fixed point of a smooth map $f: B \to B$ and $W^u_{loc}(x)$ be the local finite dimensional unstable manifold of x. The unstable set is the union $W^u(x) = \bigcup_{n\geq 0} f^n(W^u_{loc}(x))$. The topological boundary $\partial W^u(x)$ of the invariant set $W^u(x)$ is defined as $\partial W^u(x) = \omega(W^u(x) \setminus \{x\})$ where $\omega(B)$ for a set B is the usual ω-limit set of B, $\omega(B) = \bigcap_{k\geq 0} \mathrm{Cl}(\bigcup_{n\geq k} f^n(B))$. It is easy to prove that this is equivalent to the set of all $y \in B$ such that $y = \lim f^{n_i}(y_i)$, $n_i \to \infty$ as $i \to \infty$, the y_i belonging to a fundamental domain $G^u(x)$ for $W^u_{loc}(x)$. It is clear that if $A(f)$ is compact, then $\partial W^u(x)$ is an invariant set.

Proposition 10.5. *Let x be a hyperbolic fixed point of a smooth map f and assume $\dim W^u_{loc}(x) < \infty$. Then $W^u(x)$ is invariant and $\mathrm{Cl}\, W^u(x) = \partial W^u(x) \cup W^u(x)$. If, in addition, $A(f)$ is compact, then $\partial W^u(x)$ and $\mathrm{Cl}\, W^u(x)$ are invariant sets.*

We say that $x = f^n(x)$ is a non-degenerate n-periodic point if n is the period of x and $1 \notin$ spectrum of $df^n(f)$.

As an application of the Implicit Function Theorem and classical results on stable and unstable manifolds, one obtains

Proposition 10.6. *Let $x = x(f)$ be a non-degenerate n-periodic point of a C^r map $f: B \to B$, $r \geq 1$. There exist neighborhoods U of x in B and $\mathcal{V}(f)$ of f in $C^r(B,B)$ such that any $g \in \mathcal{V}(f)$ has in U only one n-periodic point $x(g)$ and no other m-periodic point with $m \leq n$. Moreover, if x is hyperbolic, the local stable and unstable manifolds depend continuously on $g \in \mathcal{V}(f)$; in particular if $W^u_{loc}(x(f))$ is finite dimensional, one has $\dim W^u_{loc}(x(f)) = \dim W^u_{loc}(x(g))$ for all $g \in \mathcal{V}(f)$.*

Proposition 10.7. Let P be a hyperbolic periodic point of a smooth map f, $\dim W^u_{loc}(P) < \infty$, and $N^u(P)$ a fundamental neighborhood for $W^u_{loc}(P)$. Then, there exists a neighborhood W of P such that

$$\bigcup_{n\geq 0} f^{-n}(N^u(P)) \cup W^s_{loc}(P) \supset W.$$

Proof: Let p be the period of P and $h = f^p$. If the proposition is not true, there exists a sequence $x_\nu \to P$ as $\nu \to \infty$ such that $x_\nu \notin W^s_{loc}(P)$ and $x_\nu \notin \bigcup_{n\geq 0} f^{-n}(N^u(P))$. Let $V = B^s \times B^u$ be a neighborhood of P considered in the construction of $N^u(P)$. Let k_ν be the first integer such that $Z_{k_\nu} = h^{k_\nu}(x_\nu) \notin V$; such a first integer does exist, otherwise $x_\nu \in W^s_{loc}(P)$. The sequence $k_\nu \to \infty$ as $\nu \to \infty$; in fact, if $k_\nu \leq M$ for all $\nu \geq 1$, since $h^{k_\nu}(P) = P$ and h^{k_ν} is continuous there exists a neighborhood $\tilde{V}$ of P, $\tilde{V} \subset V$ such that $h^{k_\nu}(\tilde{V}) \subset V$ for all $k_\nu \leq M$ which is absurd because the $x_\nu \in \tilde{V}$ for all $\nu \geq \nu_0$ imply $h^{k_\nu}(x_\nu) \in V$ giving a contradiction.

We may assume the neighborhood V is chosen such that $h(x_s,x_u) = (L_s x_s + \Phi_s(x_s,x_u), L_u x_u + \Phi_u(x_s,x_u))$ verifies $\|L_s\|, \|L_u^{-1}\| < a < 1$, $\left|\frac{\partial \Phi_i}{\partial x_j}\right| \leq k$, $a+k < 1$, $i,j = u,s$. Since $x_\nu \to P$, there exists ν_1 such that for all $\nu \geq \nu_1$ one has $\|x_\nu\| \leq \frac{1}{\nu}$ and if $h(x_s,x_u) = (\bar{x}_s,\bar{x}_u)$, $\bar{x}_s = B^s$, $\bar{x}_u \in B^u$, one obtains, also, $\|\bar{x}_s\| = \|L_s x_s + \Phi_s(x_s,x_u)\| \leq a(\frac{1}{\nu}) + k(\frac{1}{\nu}) < \frac{1}{\nu}$. The canonical projections of $h^{k_\nu-1}(x_\nu)$ on B^u and B^s are $\pi_u(h^{k_\nu-1}(x_\nu))$ and $\pi_s(h^{k_\nu-1}(x_\nu))$, respectively; since $\mathrm{Cl}\, B^u$ is compact, there is a limit point Z for $\pi_u(h^{k_\nu-1}(x_\nu))$, that is, at least for a subsequence one has $\lim_{\nu\to\infty} \pi_u(h^{k_\nu-1}(x_\nu)) = Z$ and $\|\pi_s(h^{k_\nu-1}(x_\nu))\| < \frac{1}{\nu}$ for all $\nu \geq \nu_1$. The above argument shows us that there exists a subsequence of $h^{k_\nu - 1}(x_\nu)$ which has $Z \in \mathrm{Cl}\, B^u$ as a limit and $h^{k_\nu}(x_\nu) = h(h^{k_\nu - 1}(x_\nu)) \to h(Z)$. That limit point Z is not P because $h(Z) \notin V$, then Z reaches $N^u(P)$ after a finite number of iterations of h and, by continuity, each $h^{k_\nu - 1}(x_\nu)$, for large ν, reaches $N^u(P)$ after a finite number of iterations of h, that is, $x_\nu \in \bigcup_{n \geq 0} f^{-n}(N^u(P))$ which is a contradiction. The proof is then, complete.

Proposition 10.8. Let P be a hyperbolic p-periodic point of a map $f \in C^r(B,B)$, $r \geq 1$, $\dim W^u_{loc}(P) < \infty$, and $N^u(P)$ a fundamental neighborhood for $W^u_{loc}(P)$. Then, there exist neighborhoods U of P in B and $\mathscr{V}(f)$ of f in $C^r(B,B)$ such that $N^u(P)$ is a fundamental neighborhood for $W^u_{loc}(P^*)$, $P^* = P^*(g)$ being the unique hyperbolic p-periodic point in U corresponding to $g \in \mathscr{V}(f)$. Moreover, there exists a neighborhood W_0 of P such that for all g in $\mathscr{V}(f)$ one has

$$\bigcup_{n \geq 0} g^{-n}(N^u(P)) \cup W^s_{loc}(P^*(g)) \supset W_0.$$

Proof: The first statement is a consequence of Proposition 10.6. Assume that the remaining statement is not true; then there exist sequences $x_\nu \to P$ and $g_\nu \to f$ such that

$$x_\nu \notin W^s_{loc}(P^*(g_\nu))$$

and

$$x_\nu \notin \bigcup_{n \geq 0} g_\nu^{-n}(N^u(P)).$$

As before, let k_ν be the first integer such that $g_\nu^{pk_\nu}(x_\nu) \notin V = B^s \times B^u$; such a first integer does exist because $x_\nu \notin W^s_{loc}(P^*(g_\nu))$. Call

$h = f^p$ and $h_\nu = g_\nu^p$; if $k_\nu \le M$ for $\nu \ge 1$, $h^{k_\nu}(P) = P$ implies the existence of $\tilde{V}$, neighborhood of P, such that $h^{k_\nu}(\tilde{V}) \subset V$; then $f^{pk_\nu}(x_\nu) \in V$ for large ν and $g_\nu \to f$ implies $g_\nu^{pk_\nu}(x_\nu) \in V$ which contradicts the definition of k_ν. We know that $g_\nu^{pk_\nu}(x_\nu) = h_\nu^{k_\nu}(x_\nu) \notin V$ but $g_\nu^{p(k_\nu-1)}(x_\nu) = h_\nu^{k_\nu-1}(x_\nu) \in V$ for all $\nu \ge 1$. The convergence $h_\nu \to h$ is in the C^1-norm then we can choose V such that $h_\nu(x_s,x_u) = (L_s^\nu x_s + \Phi_s^\nu(x_s,x_u), L_u^\nu x_s + \Phi_u^\nu(x_s,x_u))$, $||L_s|| < a < 1$, $\left|\frac{\partial\Phi_s}{\partial x_j}\right| \le k$, $0 < a+k < 1$ and given $\varepsilon > 0$, $\exists \nu_0$ such that $\nu \ge \nu_0$ implies $||L_s - L_s^\nu|| < \varepsilon$ and $\left|\frac{\partial\Phi_s^\nu}{\partial x_j} - \frac{\partial\Phi_s}{\partial x_j}\right| < \varepsilon$, $j = u,s$. It follows that for a suitable $\varepsilon > 0$, $||L_s^\nu|| \le a + \frac{\varepsilon}{2} < 1$ and $\left|\frac{\partial\Phi_s^\nu}{\partial x_j}\right| < k + \frac{\varepsilon}{2}$, $(a + \frac{\varepsilon}{2}) + (k + \frac{\varepsilon}{2}) = (a+k+\varepsilon) < 1$. The same argument used in the last Proposition 10.7 shows that $h_\nu^{k_\nu-1}(x_\nu) \to \overline{Z}$. If $\overline{Z} = P$, $h(\overline{Z}) = h(P) = P = h(h_\nu^{k_\nu-1}(x_\nu))$ and since $h_\nu \to h$, $|h_\nu(h_\nu^{k_\nu-1}(x_\nu)) - h(h_\nu^{k_\nu-1}(x_\nu))| < \varepsilon$ that is $|h_\nu^{k_\nu}(x_\nu) - P| < \varepsilon$ which is absurd since $h_\nu^{k_\nu}(x_\nu) \notin V$. Since $\overline{Z} \ne P$ and $\overline{Z} \in W^u_{loc}(P)$, with a finite number of iterations of $\overline{Z}$ by $h/W^u_{loc}(P)$ or its inverse one reaches $N^u(P)$ and with the same number, for large ν, x_ν reaches $N^u(P)$ by using h_ν and $x_\nu \in \bigcup_{n\ge 0} g_\nu^{-n}(N^u(P))$ which is a contradiction. The proof is complete.

To state dual results corresponding to Propositions 10.7 and 10.8 we need to assume reversibility for f and some compactness hypothesis in the set of maps to be considered.

Proposition 10.9. *Let P be a hyperbolic periodic point of a C^0-reversible and smooth map f such that $A(f)$ is compact; let $N^s(P)$ be a fundamental neighborhood for $W^s_{loc}(P)$. Then, there exists a neighborhood W of P such that*

$$\bigcup_{n\geq 0} f^n(N^s(P)) \cup W^u_{loc}(P) \supset W \cap A(f),$$

Proof: Let p be the period of P $(f^p(P) = P)$ and V the neighborhood used in the construction of $N^s(P)$. If Proposition 10.9 is not true, there exists a sequence $x_\nu \to P$, $x_\nu \in A(f)$, $x_\nu \notin W^u_{loc}(P)$ and $x_\nu \notin \bigcup_{n\geq 0} f^n(N^s(P))$. Each x_ν defines a unique sequence $(x_\nu = x^0_\nu, x^1_\nu, x^2_\nu, \ldots) \in A(f)$, $f^p(x^i_\nu) = x^{i-1}_\nu$, $i \geq 1$. Let k_ν be the first integer such that $x^{k_\nu}_\nu \notin V$ (if $x^i_\nu \in V$ for all i, $x_\nu \in W^u_{loc}(P)$). See that $k_\nu \to \infty$ as $\nu \to \infty$; if $k_\nu \leq M$ let $\tilde{f} = f/A(f)$ be the homeomorphism obtained restricting f to $A(f)$, $\tilde{f}^{-pk_\nu}(P) = P$, and given $V \cap A(f)$, there exists $\tilde{V} = \tilde{V}(P)$ such that $\tilde{f}^{-pk_\nu}(\tilde{V} \cap A(f)) = V \cap A(f)$ and for large $\nu \geq \nu_0$, $x_\nu \in \tilde{V} \cap A(f)$ then $\tilde{f}^{-pk_\nu}(x_\nu) \in V \cap A(f)$ which is a contradiction. We have limit points $x^{k_\nu - 1}_\nu \to x$, $x^{k_\nu}_\nu \to y$, $\tilde{f}^p(y) = x$ and $x \neq P$ (if $x = P \Rightarrow y = P$ (contradiction since $y \notin V$)). It is easy to see that $x \in A(f) \cap W^s_{loc}(P)$ since there exists a sequence $x, f^p(x), f^{2p}(x), \ldots$, constructed using $x^{k_\nu - 1}_\nu$, $x^{k_\nu - 2}_\nu$, and so on. With a finite number of iterations by $\tilde{f}^p$, x reaches $N^s(P)$. Since $x^{k_\nu - 1}_\nu \to x$ the same happens with $x^{k_\nu - 1}_\nu$ and $x_\nu \in \bigcup_{n\geq 0} f^n(N^s(P))$ and obtain a contradiction. The proof is complete.

Consider now a (not unique) topological subspace $KC^r(B,B)$ as defined in the beginning of the section.

Proposition 10.10. Let P be a hyperbolic p-periodic point of a map $f \in KC^r(B,B)$, $r \geq 1$, and $N^s(P)$ a fundamental neighborhood for $W^s_{loc}(P)$. Then, there exist neighborhoods U of P in B and $\mathcal{V}(f)$ of f in $KC^r(B,B)$ such that $N^s(P)$ is a fundamental neighborhood for $W^s_{loc}(P^*)$,

$P^* = P^*(g)$ _being the unique hyperbolic_ p-_periodic point in_ U _corresponding to_ $g \in \mathcal{V}(f)$. _Moreover, there exists a neighborhood_ W_0 _of_ P _such that for all_ $g \in \mathcal{V}(f)$ _one has_

$$\bigcup_{n \geq 0} g^n(N^s(P)) \cup W^u_{loc}(P^*(g)) \supset W_0 \cap A(g).$$

Proof: The first statement follows from Proposition 10.6. If the second statement is not true there exist sequences $x_\nu \to P$ and $g_\nu \to f$ such that $x_\nu \in A(g_\nu)$,

$$x_\nu \notin W^u_{loc}(P^*(g_\nu)) \quad \text{and} \quad x_\nu \notin \bigcup_{n \geq 0} g^n_\nu(N^s(P)).$$

Following the same arguments as in the proof of Proposition 10.9, each x_ν defines a unique sequence

$$(x_\nu = x^0_\nu, x^1_\nu, x^2_\nu, \ldots) \in A(g_\nu), \quad g^p_\nu(x^i_\nu) = x^{i-1}_\nu, \quad i \geq 1.$$

Let k_ν be the first integer such that $x^{k_\nu}_\nu \notin V$, V being a neighborhood of P used in the construction of $N^s(P)$ (if $x^i_\nu \in V$ for all $i \geq 0$, $x_\nu \in W^u_{loc}(P^*(g_\nu))$). The sequence g_ν may be chosen in order to obtain $A(g_\nu)$ in a $(1/\nu)$-neighborhood of $A(f)$, then $x^{k_\nu}_\nu$ approaches $A(f)$ as $\nu \to \infty$. Since $A(f)$ is compact there exists a sequence $y_\nu \in A(f)$, each y_ν giving the minimum for the distances between the $x^{k_\nu}_\nu$ and $A(f)$. The sequence y_ν has a limit point $y \in A(f)$ and it is clear that $x^{k_\nu}_\nu \to y$ as $\nu \to \infty$, then $y \notin V$. See that $k_\nu \to \infty$ as $\nu \to \infty$ (if $1 \leq k_\nu \leq M$ and since $x^{k_\nu - 1}_\nu \to f^p(y)$, $x^{k_\nu - 2}_\nu \to f^{2p}(y)$, etc., one obtains $f^{Mp}(y) = P$ which implies $y = P \in V$ - contradiction) then $P \in \omega(y)$ with respect to f^p and $y \in A(f) \cap W^s_{loc}(P)$, $y \neq P$. With a finite number of iterations of y by

$f^p/A(f)$ one reaches $N^s(P)$ and with the same number, for large ν, $x_\nu^{k_\nu}$ reaches $N^s(P)$ using g_ν^p, then $x_\nu \in \bigcup_{n \geq 0} g_\nu^n(N^s(P))$, which is a contradiction. The proof is complete.

<u>Proposition 10.11.</u> <u>Assume it is given a topological subspace</u> S <u>of</u> $C^r(B,B)$ <u>such that any</u> $f \in S$ <u>is reversible, has</u> $A(f)$ <u>compact and admits a neighborhood</u> $\mathscr{V}(f)$ <u>in</u> S <u>such that</u> $\bigcup_{g \in \mathscr{V}(f)} A(g)$ <u>is relatively compact. Then</u> S <u>has the properties of a</u> $KC^r(B,B)$.

<u>Proof:</u> If the proposition were not true, there would exist a neighborhood U_0 of $A(f)$, a sequence $f_\nu \to f$, $f_\nu \in S$ and points $x_\nu \in A(f_\nu)$ such that $x_\nu \notin U_0$. But the elements of the sequence (x_ν) eventually belong to $\bigcup_{g \in \mathscr{V}(f)} A(g)$ for a suitable neighborhood $\mathscr{V}(f)$ of f in S; so (x_ν) has a limit point x^0. Since $x_\nu \in A(f_\nu)$ there exists a sequence

$$(x_\nu = x_\nu^0, x_\nu^1, x_\nu^2, \ldots, x_\nu^i, \ldots) \in A(f_\nu)$$

such that $f_\nu(x_\nu^i) = x_\nu^{i-1}$ for $i \geq 1$. Choosing an appropriate subsequence of the (x_ν^0), hence the (x_ν^i) for each i, one obtains a sequence of limit points

$$(x^0, x^1, x^2, \ldots), \quad x_\nu^i \to x^i, \quad i \geq 0.$$

But $||f(x^1)-x^0|| \leq ||f(x^1)-f(x_\nu^1)|| + ||f(x_\nu^1)-f_\nu(x_\nu^1)|| + ||x_\nu^0-x^0||$, that is, $f(x^1) = x^0$; analogously $f(x^i) = x^{i-1}$, $i \geq 2$, and $x^0 \in A(f) - U_0 = \Phi$ which is a contradiction.

<u>Remark 10.12.</u> Let M be a compact manifold and $B = C^0(I,M)$. Let $S \subset C^r(B,B)$ be the set of all analytic flow maps f_F of analytic (RFDE) $F \in \mathscr{X}^r$ defined on the manifold M. Then S is a particular $KC^r(B,B)$,

$r \geq 1$. The reversibility of f_F follows from the analiticity of F, and each f_F being compact implies $A(f_F)$ is a compact set. The "continuity" of $A(f_F)$ with respect to f_F follows from Proposition 10.11.

In fact, the map $f: F \in \mathscr{X}^r \to f_F \in C^r(B,B)$ is well defined, continuous and injective. Moreover, $\mathscr{X}^r$ is homeomorphic to $f(\mathscr{X}^r)$ with the relative topology since $f_{F_\nu} \to f_F$ (in the topology of $C^r(B,B)$) implies $F_\nu \to F$ in $\mathscr{X}^r$. By Arzela's theorem and the above homeomorphism we see that the topological subspaces $\tilde{S} = f(\mathscr{X}^r)$ and S satisfy the hypothesis of Proposition 10.11.

The choice of the class $KC^r(B,B)$ depends on the case we are studying; for instance, maps arising from retarded functional differential equations, neutral functional differential equations, semi-linear parabolic equations, hyperbolic equations, can be considered. In each case we need to assume the appropriate hypothesis on the equations in order to obtain the compactness properties of $KC^r(B,B)$.

<u>Proposition 10.13</u>. <u>Let</u> $P,Q \in Per(f)$ <u>be distinct hyperbolic periodic points of a reversible map</u> f <u>such that</u> $\dim W^u_{loc}(P)$, $\dim W^u_{loc}(Q) < \infty$. <u>If</u> $A(f)$ <u>is compact and</u> $Cl\, W^u(Q) \cap W^u(P) \neq \Phi$, <u>then there exists</u> $x \in Cl\, W^u(Q) \cap W^s_{loc}(P)$ <u>such that</u> $x \notin \mathscr{O}(P)$.

<u>Proof</u>: From the hypothesis it follows that $P \in Cl\, W^u(Q)$ so there exists a sequence $Z_i = f^{n_i}(y_i) \to P$ with $n_i \to \infty$ as $i \to \infty$, $y_i \in G^u(Q)$. The points P and Q are fixed points of the power $g = f^{[p,q]}$, p and q being the periods of P and Q and $[p,q]$ its least common multiple. Since Q is a limit point of a sequence $(Z_i = Z_i^0, Z_i^1, Z_i^2, \ldots) \in W^u(Q)$, $g(Z_i^\nu) = Z_i^{\nu-1}$, $\nu \geq 1$, there exists a fixed integer k_i such that $Z_i^{k_i} \notin \overline{U}_0$, $\overline{U}_0$ being a suitable bounded and closed neighborhood of P, chosen together

with closed neighborhoods $\overline{U}_n$ of $f^n(P)$, $1 \leq n \leq p-1$, satisfying the condition $g(\overline{U}_i) \cap \overline{U}_j = \Phi$, $0 \leq i \neq j \leq p-1$.

Since $Z_i^{k_i} \in W^u(Q) \subset A(f)$ then the sequence $Z_i^{k_i}$ has a limit point x; but $Z_i^{k_i} \notin \text{int } \overline{U}_0$, that is, $Z_i^{k_i} \in A(f) - \text{int } \overline{U}_0$ which is closed, then $x \in A(f) - \text{int } \overline{U}_0$ and $x \neq P$.

We remark now that $k_i \to \infty$ as $i \to \infty$ because if $k_i \leq M$ for all $i \geq 1$, there exists a neighborhood $\tilde{V}$ of P such that $g^{-k_i}(\tilde{V}) \subset \overline{U}_0$ for all $k_i \leq M$. But $Z_i \to P$ implies $Z_i \in \tilde{V}$ for large i and since $g^{k_i}(Z_i^{k_i}) = Z_i$ one obtains $Z_i^{k_i} \in \overline{U}_0$ which is a contradiction.

Given $\ell \geq 1$, $g^\ell(x) \in \overline{U}_0$; in fact, for large i, $k_i > \ell$ and $g^\ell(x) = g^\ell(\lim Z_i^{k_i}) = \lim g^\ell(Z_i^{k_i}) = \lim(Z_i^{k_i-\ell}) \in \overline{U}_0$, then $x \in \text{Cl } W^u(Q) \cap W^s_{loc}(P)$. Finally, since $g(x) \in \overline{U}_0$, $x \notin \mathcal{O}(P) - \{P\}$; otherwise $x \in \overline{U}_j$ for some $0 < j \leq p-1$ which implies $g(x) = x \in \overline{U}_j \cap \overline{U}_0 \neq \Phi$.

<u>Definition 10.14</u>. Let f be an element of the topological space $KC^r(B,B)$, $r \geq 1$. We say that f is a <u>Morse-Smale map</u> ($f \in MS$) if:

1) $\Omega(f)$ is <u>finite</u> (then $\Omega(f) = \text{Per}(f)$).
2) If $P \in \text{Per}(f)$, P is <u>hyperbolic</u> and $\dim W^u(P) < \infty$.
3) If P and Q belong to Per(f) then $W^u(Q) \pitchfork W^s_{loc}(P)$ ($\pitchfork$ means <u>transversal</u>).

<u>Remark</u>. It is clear that if f is C^0-reversible one has $\Omega(f)$ finite implies $\Omega(f) = \text{Per}(f)$. In fact, $\Omega(f)$ is invariant, then the negative orbit of $x_0 \in \Omega(f) - \text{Per}(f)$ has an infinite number of points otherwise $x_0 \in \text{Per}(f)$. This is a contradiction. But, even without assuming C^0-reversibility one has: $\Omega(f)$ finite implies $\Omega(f) = \text{Per}(f)$. For otherwise there exists $x_0 \in \Omega(f) - \text{Per}(f)$ and $x_i = f^i(x_0)$ ($i \geq 0$) are in $\Omega(f)$

$(f(\Omega(f)) \subset \Omega(f)$, always, by continuity of f). Since $\Omega(f)$ is finite, the x_i $(0 \leq i < m)$ are distinct but $x_m \in \{x_0, x_1, \ldots, x_{m-1}\}$. Since $x_0 \notin \mathrm{Per}(f)$, $x_m \neq x_0$ and $x_m = x_{m-p}$ for some $p \in \{1, \ldots, m-1\}$ and then $\{f^i(x_0): i \geq m-p\} = \{x_{m-p}, \ldots, x_{m-1}\}$ is bounded away from x_0 and $x_0 \notin \Omega(f)$ which is a contradiction.

As a corollary of the local λ-lemma (see Proposition 10.4) one can easily prove the following:

Proposition 10.15 (global λ-lemma). Let $f: B \to B$ be a smooth reversible map and assume $A(f)$ is compact. Let $W^u(P)$ be the global unstable manifold of a hyperbolic fixed point P, $\dim W^u(P) = r$, and $N \subset A(f)$ be an injectively immersed invariant submanifold of B with a point q of transversal intersection with $W^s_{loc}(P)$. Then, for any given cell neighborhood B^r imbedded in $W^u(P)$, centered in P, and any $\varepsilon > 0$, there exists one r-cell in N, ε - C^1 close to B^r.

Proof: The reversibility of f ables us to define the global unstable manifold $W^u(P) \subset A(f)$ and $f/A(f)$ is a homeomorphism. The proof follows from the local λ-lemma.

Remark. We don't need N imbedded if $N = \bigcup_{k \geq 0} N_k$, $N_0 \subset N_1 \subset \ldots$, with each N_k imbedded.

Corollary 10.16. Let $P_i \in \mathrm{Per}(f)$, $i = 1,2,3$, be hyperbolic points. If $W^u(P_1)$ and $W^s_{loc}(P_2)$, $W^u(P_2)$ and $W^s_{loc}(P_3)$ have $Q_1, Q_2 \notin \mathrm{Per}(f)$ of transversal intersections then $W^u(P_1)$ and $W^s_{loc}(P_3)$ also have a point $Q_3 \notin \mathrm{Per}(f)$ of transversal intersection.

Corollary 10.17. Let $P \in \mathrm{Per}(f)$ be hyperbolic. If $W^u(P)$ meets $W^s_{loc}(P)$ in a point $Q \notin \mathcal{O}(P)$ of transversal intersection, then $\Omega(f)$ is not finite.

Let us introduce now the set MR of all elements of the topological space $KC^r(B,B)$, $r \geq 1$, such that

1) $\Omega(f)$ is finite (then $\Omega(f) = Per(f)$).

2) $P \in Per(f) \Rightarrow P$ is hyperbolic and $\dim W^u(P) < \infty$.

3) If $P,Q \in Per(f)$ and $W^u(P) \cap W^s_{loc}(Q) \neq \Phi$ then there exists a point of transversal intersection.

It is clear that $MS \subset MR$ and if $f \in MR$, $A(f)$ is the union of all unstable manifolds of $P \in Per(f)$.

Proposition 10.18. If $f \in MR$, there exist in $Per(f)$ at least one sink and at least one source. Moreover, $A(f) = \bigcup_{P \in Per(f)} W^u(P)$.

Note. It is possible the source and the sink are identical, case in which $A(f)$ is a single point.

Proof: If there are no sources in $Per(f) = \Omega(f)$ then there exists a cycle with transversal intersections and unstable manifolds with the same dimension. Using the global λ-lemma and their corollaries one concludes that $\Omega(f)$ is not finite. The same argument shows the existence of a sink.

Proposition 10.19. Let $f \in MR$ and $P,Q \in Per(f)$ such that $P \neq Q$ and $Cl\, W^u(Q) \cap W^u(P) \neq \Phi$. Then there exists a sequence $P_1,P_2,\ldots,P_n \in Per(f)$, $P_1 = P$, $P_n = Q$, such that

$$W^u(P_{i+1}) \cap W^s_{loc}(P_i) \neq \Phi, \quad 1 \leq i \leq n-1.$$

Proof: We start with some remarks:

a) If $x \in Cl\, W^u(Q) \cap W^u(P)$, x is assumed to be in $W^u_{loc}(P)$.

b) $Cl\, W^u(Q) \cap W^u(P) \neq \Phi$ if and only if $\partial W^u(Q) \cap W^u(P) \neq \Phi$.

c) P cannot be a source $(W^s_{loc}(P) \cap A(f) = \{P\})$.

In fact, the Proposition 10.13 implies that $Cl\, W^u(Q) \cap W^u(P) \neq \Phi \Rightarrow W^s_{loc}(P) \cap A(f) \neq \{P\}$.

d) If P is a sink $(W^u_{loc}(x) = \{x\})$ it is enough to define

$P = P_1$ and $Q = P_2$.

Finally, P is a saddle, then by Proposition 10.13 there exists $x \in Cl\, W^u(Q) \cap W^s_{loc}(P)$ and $x \notin \mathcal{O}(P)$; we may assume

$$x \in \partial W^u(Q) \cap W^s_{loc}(P)$$

otherwise we are done. But $\partial W^u(Q) \subset A(f)$ then $x \in A(f)$, that is, $x \in W^u(P_2)$ for some $P_2 \in Per(f)$ which implies

$$W^u(P_2) \cap W^s_{loc}(P) \neq \Phi \quad \text{and} \quad W^u(P_2) \cap \partial W^u(Q) \neq \Phi.$$

If $P_2 = Q$ the proposition is proved. If $P_2 \neq Q$ we repeat the argument and get the sequence $(P_1 = P, P_2, P_3, \ldots)$. Remark that in this sequence $P_i \neq P_j$ otherwise $\Phi \neq W^u(P_i) \cap W^s_{loc}(P_i) \neq \mathcal{O}(P_i)$ and $\Omega(f)$ is not finite by Corollary 10.17. Since $\Omega(f)$ is finite we reach the given point Q.

Proposition 10.20. Let $f \in MR$. Then for each $P \in Per(f)$, $W^u(P)$ is imbedded in B. In particular, f as a map from $W^u(P)$ into itself is differentiable.

Proof: If $W^u(P)$ is not imbedded we have $\partial W^u(P) \cap W^u(P) \neq \Phi$ and then there exists $x \in W^u(P) \cap W^s_{loc}(P)$, $x \notin \mathcal{O}(P)$, with transversality ($f \in MR$), then $\Omega(f)$ is not finite.

We introduce in the set of orbits of periodic points a partial order using the following definition:

Definition 10.21. Let $f \in MR$ and $P, Q \in Per(f)$. Then $\mathcal{O}(P) \leq \mathcal{O}(Q)$ if $C\ell\, W^u(\mathcal{O}(Q)) \cap W^u(\mathcal{O}(P)) \neq \varphi$. Here, $W^u(\mathcal{O}(Q) = \cup_{x \in \mathcal{O}(Q)} W^u(x)$.

The above definition does not depend on the choice of the particular representatives of $\mathcal{O}(P)$ and $\mathcal{O}(Q)$. If $P_1 \in \mathcal{O}(P)$ and $Q_1 \in \mathcal{O}(Q)$ we see that $Cl\, W^u(Q) \cap W^u(P) \neq \Phi$ if and only if $W^u(Q) \cap W^s_{loc}(P) \neq \Phi$, if and only if $W^u(Q_1) \cap W^s_{loc}(P_1) \neq \Phi$. The relation $\mathcal{O}(P) \leq \mathcal{O}(Q)$ is obviously reflexive and transitive by using the global λ-lemma and their corollaries. Finally if $W^u(Q) \cap W^s_{loc}(P) \neq \Phi$ and $W^u(P) \cap W^s_{loc}(Q) \neq \Phi$ for $Q \notin \mathcal{O}(P)$ we obtain a kind of cycle and the global λ-lemma shows that $\Omega(f)$ is infinite which is a contradiction. Then, $\mathcal{O}(P) = \mathcal{O}(Q)$ and $\leq$ is a partial order.

The set of orbits of all periodic points of a map $f \in MR$ together with the above defined partial order is called the phase diagram $D(f)$ of f. For $P, Q \in Per(f)$, a chain connecting Q to P in the phase diagram of f is a sequence $P_0, \ldots, P_n$ with $P_i \in Per(f)$, $P_i \notin \mathcal{O}(P_{i+1})$, $P_1 = P$ and $P_n = Q$, such that $W^u(P_{i+1}) \cap W^s_{loc}(P_i) \neq \Phi$. The integer n is the length of the chain. Q is said to have k-behavior relative to P (write $beh(Q|P) = k$) if the maximum length of chains connecting Q to P is $k \in N$; complete the definition by setting $beh(Q|P) = 0$ iff $W^u(Q) \cap W^s_{loc}(P) = \Phi$. If $Q \in \mathcal{O}(P)$ then $beh(Q|P) = 0$ but not conversely because if P, Q are fixed points and sinks we have $beh(Q|P) = 0$ and $Q \notin \mathcal{O}(P)$. It is also clear that for distinct orbits $\mathcal{O}(P) \leq \mathcal{O}(Q)$ implies $beh(Q|P) > 0$.

We will show that given $f \in MR$, there is a neighborhood $\mathcal{V}(f)$ of f in $KC^r(B,B)$ such that $g \in \mathcal{V}(f)$ implies $g \in MR$ and there is an <u>isomorphism between phase diagrams</u>, that is, a bijection $\rho(g)\colon D(f) \to D(g)$ between the phase diagrams of f and g which is ordering and indices preserving, that means: $P,Q \in Per(f)$, $\mathcal{O}(P) \leq \mathcal{O}(Q)$, implies $\mathcal{O}(\rho(g)P) \leq \mathcal{O}(\rho(g)Q)$ and $\dim W^u(P) = \dim W^u(\rho(g)P)$.

Since $f \in MR$, by Proposition 10.6 each $g \in \mathcal{V}_1(f)$ defines a map

$$\rho(g): Per(f) \to Per(g) \subset \Omega(g)$$
$$P \longrightarrow P^* = \rho(g)P.$$

We will construct neighborhoods V of $A(f)$ and $\mathcal{V}(f)$ of f such that

$$\Omega(g) \cap V = \rho(g)[Per(f) \cap V]$$

for all $g \in \mathcal{V}(f)$. We will proceed by induction on the phase diagram of f.

For each sink S_i of f, choose a neighborhood $V_0(S_i) \subset W^s_{loc}(S_i)$ and $\varepsilon_0(S_i) > 0$ such that if $|g-f|_r < \varepsilon_0(S_i)$ then $V_0(S_i) \subset W^s_{loc}(S_i^*)$, where $S_i^* = \rho(g)S_i$. Let $V_0 = \cup V_0(S_i)$ and $\varepsilon_0 = \min\{\varepsilon_0(S_i) \mid S_i$ is a sink of $f\}$. In V_0 we trivially have $\Omega(g) \cap V_0 = \rho(g)[Per(f) \cap V_0]$ for all $|g-f|_r < \varepsilon_0$. If, now, S is a saddle near sinks ($beh(S|S_i) \leq 1$ for all sinks S_i), by the compactness of the fundamental domain $G^u(S)$, there exist n_0 and a fundamental neighborhood $N^u(S)$ such that given $x \in N^u(S)$, $f^n(x) \in V_0$ for some $n \leq n_0$. The same happens with g near f; by Proposition 10.8, $\bigcup_{n>0} g^{-n}(N^u(S)) \cup W^s_{loc}(S^*)$ contains a neighborhood $U_1(S)$ of S in B, for all g belonging to a suitable $\varepsilon_1(S)$-neighborhood of f in $KC^r(B,B)$. Consider $V_1(S) = V_0 \cup [\bigcup_{n=1}^{n_0} f^{-n}(V_0)] \cup U_1(S)$ and $\varepsilon_1(S)$ for each saddle S near sinks and finally $V_1 = \cup V_1(S)$ and $\varepsilon_1 = \min\{\varepsilon_1(S)\}$ for all saddles near sinks. In V_1 we have

$$\Omega(g) \cap V_1 = \rho(g)[Per(f) \cap V_1].$$

By induction, assume now that we have constructed V_k, ε_k corresponding to the points in $\mathrm{Per}(f)$ whose behavior with respect to sinks of f is $\leq k$, so that $\Omega(g) \cap V_k = \rho(g)[\mathrm{Per}(f) \cap V_k]$ for $|g-f|_r < \varepsilon_k$. Let P_{k+1} be a point next to these in the phase-diagram of f. Again, by the compacity of $G^u(P_{k+1})$ there exists $n_1(P_{k+1})$ such that $f^n(x) \in V_k$ for all $x \in G^u(P_{k+1})$ and some $1 \leq n \leq n_1(P_{k+1})$. Using inverse images of V_k by f one defines $N^u(P_{k+1})$ and $\varepsilon_{k+1}(P_{k+1})$; for $|g-f|_r < \varepsilon_{k+1}(P_{k+1})$ the same happens with g. Use again Proposition 10.8 to obtain $U_{k+1}(P_{k+1})$ = neighborhood of $P_{k+1} \subset W^s_{loc}(P^*_{k+1}) \cup \bigcup_{n \geq 0} g^{-n}(N^u(P_{k+1}))$. Define $U_{k+1} = \cup U_{k+1}(P_{k+1})$ and $\varepsilon_{k+1} = \min\{\varepsilon_{k+1}(P_{k+1})\}$, $n_1 = \max\{n_1(P_{k+1})\}$; finally

$$V_{k+1} = V_k \cup \left[\bigcup_{n=1}^{n_1} f^{-n}(V_k)\right] \cup U_{k+1}$$

and in V_{k+1} we have

$$\Omega(g) \cap V_{k+1} = \rho(g)[\mathrm{Per}(f) \cap V_{k+1}]$$

for all $|g-f|_r < \varepsilon_{k+1}$. The induction is complete. Remark that in V_{k+1} there are no other non-wandering points besides $P_i \in \mathrm{Per}(f)$ and the corresponding P^*_i of g. The procedure reaches the sources and we define the above mentioned neighborhoods V of $A(f)$ and $\mathcal{V}(f)$ of f such that

$$\Omega(g) \cap V = \rho(g)[\mathrm{Per}(f) \cap V]$$

for all $g \in \mathcal{V}(f)$. But $f \in KC^r(B,B)$ and we reduce $\mathcal{V}(f)$, if necessary, and obtain $A(g) \subset V$ for all $g \in \mathcal{V}(f)$. Then, since $\Omega(g) \subset A(g) \subset V$, it follows that $\Omega(g) = \mathrm{Per}(g)$ for all $g \in \mathcal{V}(f)$ and we have finished the proof of the following:

Theorem 10.22. The set MR is open in $KC^r(B,B)$, $r \geq 1$. Moreover, if $f \in MR$ there is a neighborhood $\mathcal{V}(f)$ of f in $KC^r(B,B)$ such that for each $g \in \mathcal{V}(f)$ the map $\rho(g)$: $Per(f) \to Per(g)$ considered above is a diagram isomorphism. In particular, f is Ω-stable.

Consider again a smooth map $f \in MR$. If $P_k, P_{k+1} \in Per(f)$ satisfy $beh(P_k|P_{k+1}) = 1$ and if $G^s(P_{k+1})$ is a fundamental domain (then compact) for $W^s_{loc}(P_{k+1})$ we have that $W^u(P_k) \cap G^s(P_{k+1})$ is also compact. In fact, if $x_\nu \to x$, $x_\nu \in W^u(P_k) \cap G^s(P_{k+1})$, it is clear that $x \in G^s(P_{k+1})$ and if $x \notin W^u(P_k)$ (then $x \in \partial W^u(P_k)$), there exists $\tilde{P} \in Per(f)$ such that $x \in W^u(\tilde{P})$, $\tilde{P} \neq P_{k+1}$ and $\tilde{P} \neq P_k$; but by Proposition 10.19 $Cl\, W^u(P_k) \cap W^u(\tilde{P}) \neq \Phi$ implies $W^u(P_k) \cap W^s_{loc}(\tilde{P}) \neq \Phi$, then $beh(P_k|P_{k+1}) > 1$ giving us a contradiction, that is, $x \in W^u(P_k)$.

The Proposition 10.10 combined with Theorem 10.22, Proposition 10.4 and the arguments of transversality of manifolds prove the following:

Proposition 10.23. Let $f \in MS$, $P \in Per(f)$ and $\dim W^u(P) = m$. Fix a cell neighborhood B^m of P in $W^u_{loc}(P)$. Given $\varepsilon > 0$, there exist neighborhoods V of P, and $\mathcal{V}(f)$ of f in $KC^r(B,B)$, $r \geq 1$, such that if for some $Q \in Per(f)$, $W^u(Q^*(g)) \cap V \neq \Phi$ then $W^u(Q^*(g)) \cap V$ is fibered by m-cells ε - C^1 close to B^m, $g \in \mathcal{V}(f)$ and $Q^*(g) = \rho(g)Q$.

From Theorem 10.22 and Proposition 10.23 we obtain the following result.

Theorem 10.24. The set MS of all r-differentiable Morse-Smale maps is open in MR (then in $KC^r(B,B)$), $r \geq 1$. Moreover, if $f \in MS$, then its phase-diagram is stable (up to a diagram isomorphism) under small C^r perturbations of f in $KC^r(B,B)$.

Remark. In proving Proposition 10.23, we really have an Unstable Foliation of $U = V \cap A(f)$ for $f \in MS$ at $P \in Per(f)$, that is, a continuous foliation $\mathcal{F}^u(P)$: $x \in U \to \mathcal{F}^u_x(P)$ such that:

a) the leaves are C^1 discs, varying continuously in the C^1 topology and $\mathcal{F}^u_P(P) = W^u(P) \cap U$,

b) each leaf $\mathcal{F}^u_x(P)$ containing $x \in U$, is contained in U,

c) $\mathcal{F}^u(P)$ is f-invariant; that is, $f(\mathcal{F}^u_x(P)) \supset \mathcal{F}^u_{f(x)}(P)$, x and $f(x)$ in U.

Moreover, using the reversibility property of the MS maps, this unstable foliation can be easily globalized through saturation by f. The same happens for g in a suitable neighborhood $\mathcal{V}(f)$ of f in MS (then in $KC^r(B,B)$).

By induction on the phase diagram of $f \in MS$ and using the global λ-lemma we easily obtain a so-called compatible system of global unstable foliations $\mathcal{F}^u(P_1), \mathcal{F}^u(P_2), \ldots, \mathcal{F}^u(P_n)$, for any maximal chain $(P_1, P_2, \ldots, P_n) \in Per(f)$, $\mathcal{O}(P_i) \leq \mathcal{O}(P_{i+1})$, $i = 1,2,\ldots,n-1$, P_1 being a source and P_n being a sink. The compatibility means that "if a leaf F of $\mathcal{F}^u(P_k)$ intersects a leaf $\tilde{F}$ of $\mathcal{F}^u(P_\ell)$, $k < \ell \leq n$, then $F \supset \tilde{F}$; moreover, the restriction of $\mathcal{F}^u(P_\ell)$ to a leaf of $\mathcal{F}^u(P_k)$ is a C^1 foliation."

In a sequel we will prove a stability theorem for Morse-Smale maps.

Definition 10.25. A map f in $KC^r(B,B)$ is A-stable if there exists a neighborhood $\mathcal{V}(f)$ of f in $KC^r(B,B)$ such that to each $g \in \mathcal{V}(f)$ one can find a homeomorphism $h = h(g)$: $A(f) \to A(g)$ satisfying the conjugacy condition $h \cdot f = g \cdot h$ on $A(f)$.

The properties of $f \in MS$, specially the reversibility of f and the compactness of $A(f)$, the finite dimensionality of the unstable manifolds $W^u(P)$, $P \in Per(f)$, the existence of compatible systems of global unstable foliations and the parametrized version of the Isotopy Extension Theorem are the main tools to be used in the proof of the next Theorem 10.27.

In order to recall the Isotopy Extension Theorem (IET) one needs some more notation.

Let N be a C^r compact manifold, $r \geq 1$ and A an open set of R^s. Let M be a C^∞ manifold with $\dim M > \dim N$. We indicate by $C^k_A(N\times A, M\times A)$ the set of C^k mappings $f: N\times A \to M\times A$ such that $\pi = \pi' \cdot f$, endowed with the C^k topology, $1 \leq k \leq r$. Here, π and π' denote the natural projections $\pi: N\times A \to A$, $\pi': M\times A \to A$. Let $\mathrm{Diff}^k_A(M\times A)$ be the set of C^k diffeomorphisms φ of $M\times A$ such that $\pi' = \pi'\cdot\varphi$, again with the C^k topology.

Lemma 10.26. (Isotopy Extension Theorem). Let $i \in C^k_A(N\times A, M\times A)$ be an imbedding and A' a compact subset of A. Given neighborhoods U of $i(N\times A)$ in $M\times A$ and V of the identity in $\mathrm{Diff}^k_A(M\times A)$, there exists a neighborhood W of i in $C^k_A(N\times A, M\times A)$ such that for each $j \in W$ there exists $\varphi \in V$ satisfying $\varphi\cdot i = j$ restricted to $N\times A'$ and $\varphi(x) = x$ for all $x \notin U$.

Theorem 10.27. Any Morse-Smale map f in $KC^r(B,B)$ is A-stable.

Proof: By Theorem 10.24 (openess) there exists a neighborhood of f in $KC^r(B,B)$ containing only Morse-Smale maps. We say, also, that if $P_k, P_{k+1} \in Per(f)$ satisfy $beh(P_k|P_{k+1}) = 1$ then $W^u(P_k) \cap G^s(P_{k+1})$ is compact. If P_1 is a source and $beh(P_1, P_{k+1}) = k$, there exists a maximal chain $(P_1, P_2, \ldots, P_{k+1})$ such that $\mathrm{beh}(P_i, P_{i+1}) = 1$, $i = 1,2,\ldots,k$. Recall that $G^s(P_{k+1}) = \mathrm{Cl}[B^s \cap A(f)] - f(B^s \cap A(f))$. Since the compact set $A(f)$ is equal to the union of all global unstable manifolds of periodic points of f (Prop. 10.18) and $\Omega(f) = \mathrm{Per}(f)$ is finite, we may assume that $B^s = B^s(P_{k+1})$ have been chosen in such a way that $A(f)$ is transversal to ∂B^s (besides being transversal to B^s). This requires explanation since ∂B_s is not generally differentiable. We may however choose B_s so ∂B_s - near $A(f)$ - is a differentiable manifold and transverse to $A(f)$. The crucial property we need is that, given $x \in \partial B_s \cap W^u(Q)$, $Q \in \mathrm{Per}(f)$, there exist $x', x'' \in W^u(Q) \cap W^s_{loc}(P_{k+1})$ arbitrarily close to x, with $x' \in B_s$, $x'' \notin \overline{B}_s$. Call $S_E = S_E(P) = \partial B^s \cap G^s(P)$; we have also $S_E = \partial B^s \cap A(f)$. In fact, $S_E \subset \partial B^s \cap \mathrm{Cl}[B^s \cap A(f)] \subset \partial B^s \cap A(f)$ trivially. For the reverse inclusion, let $x \in \partial B^s \cap A(f)$; since $f(B^s) \subset B^s$ and $x \notin B^s$, $x \notin f(B^s \cap A(f))$ while $x \in \mathrm{Cl}\, B^s \cap A(f)$, so we only need to prove $x \in \mathrm{Cl}[B^s \cap A(f)]$. For some Q, $x \in W^u(Q) \cap \partial B^s$, and these meet transversally so there exist $x' \in W^u(Q) \cap B^s$ arbitrarily close to x, i.e., $x \in \mathrm{Cl}[W^u(Q) \cap B^s] \subset \mathrm{Cl}[A(f) \cap B^s]$. We have incidentally proved $\mathrm{Cl}[A(f) \cap B^s] = A(f) \cap \mathrm{Cl}\, B^s$, which will be needed later. Remark finally that, using the relative topology of $A(f) \cap W^s_{loc}(P)$, we have $\partial G^s(P) = S_E \cup S_I$, $S_I = S_I(P) = f(S_E)$, "∂" relative to $A(f) \cap W^s_{loc}(P)$. In fact, $G^s(P) = (B^s \cup \partial B^s) \cap A(f) - f(B^s \cap A(f)) = [B^s \cap A(f) - f(B^s \cap A(f))] \cup S_E = [(\mathrm{Int}\, G^s(P)) \cup f(S_E)] \cup S_E = \mathrm{Int}\, G^s(P) \cup (S_E \cup S_I)$.

The stable set $W^s(P)$ is the set of all points $x \in B$ such that $\omega(x) = \{P\}$. Any point $z \in W^s(P) \cap A(f)$ reaches $G^s(P) - S_I(P)$ after a finite number of iterations of $\overline{f}$ or $(\overline{f})^{-1}$, $\overline{f} = f/A(f)$.

Given any bounded imbedded disc $D \subset W^u(P) = W^u(P;f)$, for g C^1-close to f there is a disc $D^* \subset W^u(P^*(g)) = W^u(P^*;g)$ C^1-close to D,

$P^* \in D^*$, where $P^* = \rho(g)P$; we say $W^u(P^*;g)$ is C^1-close to $W^u(P;f)$ "on compact sets."

Let P_2 be a periodic point of f with behavior ≤ 1 with respect to sources and consider a pair (P_1,P_2) such that P_1 is a source and $\mathrm{beh}(P_1/P_2) = 1$. The manifolds $W^u(P_1;f)$ and $W^u(P_1^*;g)$ are C^1-close on compact sets and let h_1' be the corresponding diffeomorphism; also $W^s_{loc}(P_2;f)$ and $W^s_{loc}(P_2^*;g)$ are C^1-close for g in a suitable neighborhood of f, $P_2^* = \rho(g)P_2$. By the implicit function theorem and the transversality conditions $W^u(P_1;f) \cap W^s_{loc}(P_2;f)$, $W^u(P_1^*;g) \cap W^s_{loc}(P_2^*;g)$, there is a well defined diffeomorphism $\overline{h}_2$ from $G^s(P_2;f) \cap W^u(P_1;f)$ into $W^s_{loc}(P_2^*;g) \cap W^u(P_1^*;g)$. Define a differentiable map h_2 from $G^s(P_2;f) \cap W^u(P_1;f)$ into $W^u(P_1^*,g)$ equal to $\overline{h}_2$ on $W^u(P_1;f) \cap S_E(P_2)$ and equal to $\tilde{h}_2 = g\cdot\overline{h}_2\cdot f^{-1}$ on $W^u(P_1;f) \cap S_I(P_2)$. To construct h_2 we use the IET (Lemma 10.26) just observing that $(h_1')^{-1}\cdot\overline{h}_2$ maps $W^u(P_1;f) \cap S_E(P_2)$ into $W^u(P_1;f)$ and $(h_1')^{-1}\cdot\tilde{h}_2 = (h_1')^{-1}\cdot(g\cdot\overline{h}_2\cdot f^{-1})$ maps $W^u(P_1;f) \cap S_I(P_2)$ into $W^u(P_1;f)$, both are near the corresponding inclusion maps and so can be extended to an imbedding of $G^s(P_2;f) \cap W^u(P_1;f)$ into $W^u(P_1;f)$. The property we obtain for h_2 is that $h_2f(x) = gh_2(x)$ for all $x \in W^u(P_1;f) \cap S_E(P_2)$; in fact, $gh_2(x) = g\overline{h}_2(x)$ and $h_2f(x) = h_2(f(x)) = \tilde{h}_2(f(x)) = g\overline{h}_2f^{-1}(f(x)) = g\overline{h}_2(x)$. This map h_2 can be extended to $z \in W^s(P_2;f) \cap W^u(P_1;f)$ since there exists a unique $n \in \mathbb{Z}$ such that

$$(\overline{f})^n(z) \in (G^s(P_2;f) - S_I(P_2)) \cap W^u(P_1;f):$$

define $h_2(z) = g^{-n}(h_2(f^n(z)))$ and $h_2(P_2) = P_2^*$.

We do the same with all sources $F_i \in \mathrm{Per}(f)$ such that $\mathrm{beh}(F_i|P_2) = 1$ and h_2 is defined on $W^s(P_2;f) \cap A(f)$. For the remaining points $\tilde{P}_2 \in \mathrm{Per}(f)$

with behavior ≤ 1 with respect to sources procede analogously and obtain h_2 defined on $W^s(\tilde{P}_2;f) \cap A(f)$ satisfying $h_2 f = gh_2$ and $h_2(\tilde{P}_2) = \tilde{P}_2^*$.

The next step is the consideration of $P_3 \in Per(f)$ with behavior ≤ 2 with respect to sources and we will construct a homeomorphism h_3 on $W^s(P_3;f) \cap A(f)$ starting with $G^s(P_3) - S_I(P_3)$. For the sources with behavior 1 relative to P_3 the procedure is equal to that above. Let now P_1 be a source in $Per(f)$ such that $beh(P_1|P_3) = 2$. We have at least one sequence $(P_1P_2P_3)$ such that $beh(P_1|P_2) = beh(P_2|P_3) = 1$. Since $beh(P_2|P_3) = 1$ we define a diffeomorphism $\bar{\bar{h}}_3$ on $G^s(P_3;f) \cap W^u(P_2;f)$ exactly as we did above with h_2. But $W^u(P_1;f)$ approaches $W^u(P_2;f)$ and it is well defined a foliation on $W^u(P_1;f)$ induced by $W^u(P_2;f)$; the same happens with $W^u(P_1^*;g)$ relatively to $W^u(P_2^*;g)$ for g near f in MS. The existence of a compatible system of global unstable foliations guarantees that $W^u(P_1;f)$ intersects $W^s_{loc}(P_3;f)$ with its leaves accumulating in the (compact) set $W^u(P_2;f) \cap G^s(P_3;f)$. To each leaf $\mathscr{F}_x$ of $W^u(P_1;f) \cap G^s(P_3;f)$ near $W^u(P_2;f) \cap G^s(P_3;f)$ corresponds a unique point $x \in W^s_{loc}(P_2;f) \cap W^u(P_1;f)$ near P_2. Using h_2 (defined in the P_2 level), to $\mathscr{F}_x$ corresponds a unique leaf $\mathscr{F}^*_{h_2(x)}$ of $W^u(P_1^*;g) \cap G^s(P_3^*;g)$. Consider the map $\bar{\bar{h}}_3$ defined on $G^s(P_3;f) \cap W^u(P_2;f)$ and use the C^1-closeness on compact sets of $W^u(P_2;f)$ with the leaves of $W^u(P_1;f)$ [respectively of $W^u(P_2^*;g)$ with the leaves $W^u(P_1^*;g)$] to obtain a diffeomorphism $i_x\colon \mathscr{F}_x \to W^u(P_2;f) \cap G^s(P_3;f)$ [respectively $i_x^*\colon \mathscr{F}^*_{h_2(x)} \to W^u(P_2^*;g) \cap G^s(P_3^*;g)$] and construct $\bar{h}_3 = (i_x^*)^{-1} \cdot \bar{\bar{h}}_3 \cdot i_x$ which is an extension of $\bar{\bar{h}}_3$ to the leaf $\mathscr{F}_x$. As before, one considers $\bar{h}_3$ locally defined on $W^u(P_1;f) \cap G^s(P_3;f) \cap S_E(P_3)$ and defines $\tilde{h}_3 = g \cdot \bar{h}_3 \cdot f^{-1}$ (locally) on

$W^u(P_1;f) \cap G^s(P_3;f) \cap S_I(P_3)$. Using again the IET to the local foliation $x \to \mathscr{F}_x$, x in a neighborhood of P_2 in $W^s_{loc}(P_2;f) \cap W^u(P_1;f)$, one obtains a continuous (local) extension h_3 of $\bar{\bar{h}}_3$ coinciding with $\bar{h}_3$ on $W^u(P_1;f) \cap G^s(P_3;f) \cap S_E(P_3)$ and with $\tilde{h}_3$ on $W^u(P_1;f) \cap G^s(P_3;f) \cap S_I(P_3)$. Notice that $W^u(P_1;f) \cap W^s_{loc}(P_3;f)$ and $W^u(P_1^*;g) \cap W^s_{loc}(P_3^*;g)$ are C^1-close on compact sets. In order to extend h_3 (defined on the leaves of $W^u(P_1;f) \cap G^s(P_3;f)$ near $W^u(P_2;f) \cap G^s(P_3;f))$ to $W^u(P_1;f) \cap G^s(P_3)$, we extract a small tubular neighborhood of $W^u(P_2;f) \cap G^s(P_3;f)$ in $\mathrm{Cl}[W^u(P_1;f) \cap W^s_{loc}(P_3;f)]$ and apply again the IET for diffeomorphisms near the identity. In this way we can continuously extend h_3 to a full neighborhood of $W^u(P_2;f) \cap G^s(P_3;f)$ so that it satisfies the conjugacy equation $h_3f = gh_3$ for points of $W^u(P_1;f) \cap G^s(P_3;f) \cap S_E(P_3)$.

We proceed, in an analogous way, with all possible sequences $(P_1,P_2',P_3) \in \mathrm{Per}(f)$ such that $\mathrm{beh}(P_1|P_2') = \mathrm{beh}(P_2'|P_3) = 1$. Consider, finally, the remaining sources $P_1' \in \mathrm{Per}(f)$ in the same conditions as P_1 and obtain a continuous h_3 defined on $G^s(P_3)$ with the equality $h_3f = gh_3$ holding on $S_E(P_3)$ and then, a continuous h_3 defined on $A(f) \cap W^s(P_3)$, $h_3(P_3) = P_3^*$, with the desired conjugacy property $h_3f = gh_3$.

The last step showed us, clearly, the full induction procedure. Assume we have constructed all maps h_k, satisfying $h_k(P_k) = P_k^*$ and $h_kf = gh_k$ on $A(f) \cap W^s(P_k)$ for all $P_k \in \mathrm{Per}(f)$ such that $\mathrm{beh}(F_i|P_k) \leq k-1$, $k \geq 3$, where the F_i are all sources of $\mathrm{Per}(f)$; let $P_{k+1} \in \mathrm{Per}(f)$ be such that $\mathrm{beh}(F_i|P_{k+1}) \leq k$ for all sources $F_i \in \mathrm{Per}(f)$. Let $(F_1,P_2,\ldots,P_k,P_{k+1})$ be a sequence such that F_1 is a source and $\mathrm{beh}(F_1|P_2) = \mathrm{beh}(P_2|P_3) = \ldots = \mathrm{beh}(P_k,P_{k+1}) = 1$. We start the construction of h_{k+1} on $W^u(P_k;f) \cap G^s(P_{k+1})$, extend locally h_{k+1} to $W^u(P_{k-1}) \cap G^s(P_{k+1})$ and by a second induction procedure extend h_{k+1} to

$W^u(P_{k-2}) \cap G^s(P_{k+1}),\dots,W^u(F_1) \cap G^s(P_{k+1})$, as we did in the case $k = 2$. Do the same with all maximal sequences $(F_1,P_2',P_3',\dots,P_{k+1})$ with F_1 and P_{k+1} fixed and, finally, with the remaining sources F_i to obtain h_{k+1} defined on $G^s(P_{k+1})$ verifying the equality $h_{k+1}f = gh_{k+1}$ defined on $S_E(P_{k+1})$. By forcing the conjugacy $h_{k+1}f = g\,h_{k+1}$ extend h_{k+1} to $A(f) \cap W^s(P_{k+1})$. The induction is complete and we reach the sinks. Since the disjoint union

$$\bigcup_{P \in Per(f)} A(f) \cap W^s(P)$$

is equal to $A(f)$ the map $H = h_2 \cup h_3 \cup \dots$ is well defined on $A(f)$, $H(P) = P^*$, and $Hf(x) = gH(x)$ for all $x \in A(f)$.

The final step is to check the continuity of $H\colon A(f) \to A(g)$. Remark, first of all, that if H is continuous in $f(x)$ then H is continuous in $x \in A(f)$; in fact let $z_i \to x$, $z_i \in A(f)$; since f is continuous, $f(z_i) \to f(x)$ and the fact that H is continuous at $f(x)$ implies $Hf(z_i) \to Hf(x)$ that is, $gH(z_i) \to gH(x)$. But $H(z_i) \in A(g)$, $H(x) \in A(g)$ and g is reversible, then, $H(z_i) \to H(x)$. Given, now, $x \in A(f)$, it is clear that $x \in A(f) \cap W^s(P_k)$ for some $P_k \in Per(f)$; it is sufficient to verify the continuity of H at the points x of a neighborhood of P_k in $A(f) \cap W^s(P_k)$. If P_k is a source or a sink the continuity is trivial. Assume P_k is a saddle and let $x_n \to x$, $x_n \in \mathcal{F}^u_{x_n}(P_k)$ and $x \in \mathcal{F}^u_x(P_k)$, $\mathcal{F}^u(P_k)$ being the global unstable foliation at P_k above considered. But, by the definition of $H = h_2 \cup h_3 \cup \dots$ and by the construction of the maps h_k, we see that the set of accumulation points of $\{H(x_n)\}$ is contained in $W^s_{loc}(P_k^*;g)$ and $H(x_n) \in \mathcal{F}^{u^*}_{h_k(x_n)}(P_k^*)$, $\mathcal{F}^{u^*}(P_k^*)$ being the global unstable foliation at P_k^*. Then $H(x_n) \to \mathscr{F}^{u*}_{h_k(x_n)}(P_k^*) \cap W^s_{loc}(P_k^*;g) = h_k(x) = H(x)$ proving the continuity of H. Similarly, H^{-1} is also continuous and the proof is complete.

Corollary 10.28. Let B be a compact manifold and $\mathrm{Diff}^r(B)$ the set of all C^r diffeomorphisms of B, $r \geq 1$. Then the Morse-Smale diffeomorphisms of $\mathrm{Diff}^r(B)$ are stable and form an open set.

Proof: Remark that $\mathrm{Diff}^r(B)$ satisfies the conditions to be a $KC^r(B,B) \subset C^r(B,B)$. In fact $A(f) = B$ for all $f \in \mathrm{Diff}^r(B)$ and the reversibility is trivial. The result follows from Theorem 10.27.

Theorem 10.29. Let $KC^r(B,B)$ be the subspace S, set of flow maps of all analytic RFDE $F \in \mathscr{X}^r$, $r \geq 1$, defined on an analytic compact manifold M. The Morse-Smale maps f of S are stable relatively to A(f) and form an open set in S.

Proof: Follows from Remark 10.12, Theorem 10.24 and Theorem 10.27.

11. Bibliographical Notes

Section 1. The abstract framework in this section was introduced in Hale [11]. Hale and Lopes [14] (see also Hale [10], Massatt [32,33]) proved the result in Section 1 that a compact dissipative α-contraction $T_f(t)$ has $A(f)$ compact. Billotti and LaSalle [3] proved the same result for point dissipative maps $T_f(t)$ which are completely continuous for $t \geq r$. For some other evolutionary systems which are special cases of the abstract framework in Section 1, see Hale [11], Massatt [34]. For partial results on question Q5, see Chernoff and Marsden [52], Hale and Scheurle [53].

Section 2. The concept of an RFDE on a manifold as well as Theorems 2.1 - 2.3 are due to Oliva [37,38]. See [7] for results on global analysis.

Section 3. Examples 3.3, 3.9 were given by Oliva [38], Example 3.8 by Oliva [40]. Example 3.12 is in Hale [9]. Example 3.13 is due to Oliva [39].

Section 4. Properties of local stable and unstable manifolds of critical points and periodic orbits can be found in Hale [9]. The first proof that G_0^k, G_1^k, $G_{3/2}^k(T)$, $G_2^k(T)$ are open in $\mathscr{X}^k(I,M)$, $k \geq 1$ was given by Oliva [37], [41]. Mallet-Paret [30] proved Theorem 4.1 even for the more general case when the Whitney topology is used. Although the proof follows the pattern that was developed in Peixoto [47] (see also Abraham and Robbin [1]), Lemmas 4.3, 4.4, 4.5 contain essential new ideas. The analyticity used in the proof of Lemma 4.5 is due to Nussbaum [36]. For Smale's version of Sard's theorem, see [49] or [1].

Section 5. For an historical discussion of the existence of maximal compact invariant sets, see Hale [9,11]. The proofs of all results also can be found there. We remark that more sophisticated results on dissipative systems have been obtained by Massatt [33].

Section 6. For other properties of Hausdorff dimension, see [20] and [19]. Mallet-Paret [28] proved the compact attractor had finite Hausdorff dimension in a separable Hilbert space. Mañé [31] proved the more general results in Theorems 6.1, 6.2, 6.3 and 6.4. The proof of Lemma 6.5 and Theorem 6.6 are due to Mallet-Paret [29]. Theorem 6.6 was stated by Kurzweil [22] for the case in which F is a delay equation. Example (6.4) is due to Oliva [42]. Example (6.5) is due to Popov [48] and is also discussed in [9]. The remark about the period module of any almost periodic solution of F follows from Cartwright [4]. Corollary 6.7 is due to Mallet-Paret [28]. Theorem 6.8, showing that smoothness is necessary for A(F) to have finite dimension, is due to Yorke [50].

Section 7. The proof of Lemma 7.1, Theorem 7.3, Lemma 7.4, Theorem 7.5 may be found in Oliva [40]. The proof of Theorem 7.2 can be found in Oliva [43]. Example 7.1 is due to Oliva [38]. The method of obtaining the estimates for the contraction property in the proof of Lemma 7.4 follows closely the computation in Kurzweil [23] and Lewowicz [27]. Theorem 7.6 was first proved by Kurzweil [22], [23], [24] where he also presented other interesting results for RFDE's near ordinary differential equations. Theorem 7.7 is due to Kurzweil [22], but the proof in the text is new. See also [25].

Section 8. The proof of Proposition 8.2 may be found in [9]. The remark on the nonexistence of a generic Hopf bifurcation for (8.1) is contained in Hale [12] . The proof that there are five equivalence classes of $A_{b,g}$ for (8.1) can be found in Hale and Rybakowski [16]. The results and historical references for Equation 8.2 can be found in Chafee and Infante [5], Henry [17] and Hale [13]. The example (8.8) on the elastic beam was studied by Ball [2]. To completely analyze A(f) with clamped ends, new techniques seem to be required. The case of hinged ends can be defined in some detail. The proof that A(f) is compact for the Navier-Stokes equation in a two dimensional domain can be found in Ladyzhenskaya [26].

Section 9. These results are due to Oliva [39]. The proof that the solutions of the Equation (9.6) are bounded and approach a limit as $t \to \infty$ follows from more general results of Cooke and Yorke [6]. For the second statement of Theorem 9.3 see also [21].

Section 10. The main results of this section are due to Oliva [43]. A reference for a theory of local stable and unstable manifolds of a hyperbolic fixed point of a C^r-map is [18]. Proposition 10.4 (local λ - lemma) is due to Palis [44]. The proof of Prop. 10.23 is a simple generalization of Lemma 1.11 of [44]. The language of unstable foliations and compatible system of unstable foliations is due to Palis and Takens [46] where we can see also the statement and references for a proof of the Isotopy Extension Theorem (Lemma 10.26). Finally, Theorems 10.24 and 10.27 applied to the special case in which B is a compact manifold yield the proof for the stability of Morse-Smale diffeomorphisms (Corollary 10.28), originally established in [44] and [45]. Theorem 10.29 gives, in some sense, the answer to a fundamental question established in [9], [42] and [11].

References

[1] Abraham, R. and J. Robbin, Transversal Mappings and Flows. Benjamin, 1967.

[2] Ball, J., Saddle point analysis for an ordinary differential equation in a Banach space, and an application to dynamic buckling of a beam. Nonlinear Elasticity (Ed. R. W. Dickey), Academic Press, New York, 1973, 93-160.

[3] Billoti, J. E. and J. P. LaSalle, Periodic dissipative processes. Bull. Amer. Math. Soc. 6 (1971), 1082-1089.

[4] Cartwright, M. L., Almost periodic differential equations and almost periodic flows, J. Differential Eqns., 5 (1962), 167-181.

[5] Chafee, N. and E. F. Infante, A bifurcation problem for a nonlinear partial differential equation of parabolic type. Applicable Analysis, 4 (1974), 17-37.

[6] Cooke, K. L. and J. A. Yorke, Some equations modelling growth processes and gonorrhea epidemics, Math. Biosci., 16 (1973), 75-101.

[7] Eells Jr., J., A setting for global analysis. Bull. Amer. Math. Soc., 72 (1966), 751-807.

[8] Fink, A. M., Almost Periodic Differential Equations, Lecture Notes in Math., vol. 337, Springer-Verlag, 1974.

[9] Hale, J. K., Theory of Functional Differential Equations, Springer-Verlag, New York, 1977.

[10] Hale, J. K., Some results on dissipative processes, in Lecture Notes in Math., vol. 799, 152-172, Springer-Verlag, 1980.

[11] Hale, J. K., Topics in Dynamic Bifurcation Theory, CBMS Regional Conference Series in Math., No. 47, Am. Math. Soc., Providence, R.I., 1981.

[12] Hale, J. K., Generic properties of an integro-differential equation, Am. J. of Math. To appear.

[13] Hale, J. K., Dynamics in parabolic equations-an example, Proceedings of the Nato Conference on Nonlinear PDE's, July, 1982.

[14] Hale, J. K. and O. Lopes, Fixed point theorems and dissipative processes. J. Differential Eqns. 13 (1973), 391-402.

[15] Hale, J. K. and P. Massatt, Asymptotic behavior of gradient-like systems. Univ. Fla. Symp. Dyn. Systems, II, Academic Press, 1982.

[16] Hale, J. K. and K. P. Rybakowski, On a gradient-like integro-differential equation, Proc. Royal Soc. Edinburgh, 92A (1982), 77-85.

[17] Henry, D., Geometric Theory of Semilinear Parabolic Equations. Lecture Notes in Math. Vol. 840, Springer-Verlag, 1981.

[18] Hirsh, M. W., Pugh, C. C. and M. Shub, Invariant Manifolds, Lecture Notes in Math., vol. 583, Springer-Verlag, 1977.

[19] Hurewicz, W. and H. Wallman, Dimension Theory, Princeton University Press, 1948.

[20] Kahane, J. P., Mésures et dimensions, in Lecture Notes in Math., vol. 565, Springer-Verlag, 1976.

[21] Kaplan, J. K. and J. A. Yorke, Ordinary differential equations which yield periodic solutions of differential delay equations, J. Math. Anal. Appl., 48 (1974), 317-325.

[22] Kurzweil, J., Global solutions of functional differential equations, in Lecture Notes in Math., vol. 144, Springer-Verlag, 1970.

[23] Kurzweil, J., Invariant manifolds I, Comm. Math. Univ. Carolinae, 11 (1970), 336-390.

[24] Kurzweil, J., Invariant manifolds for flows, in Differential Equations and Dynamical Systems, 431-468, Eds. Academic Press, 1967.

[25] Kurzweil, J., Small delays don't matter, in Lectures Notes in Math., vol. 206, 47-49, Springer-Verlag, New York, 1971.

[26] Ladyzhenskaya, O. A., A dynamical system generated by the Navier-Stokes equation. J. Soviet Math. 3 (1975), 458-479.

[27] Lewowicz, J., Stability properties of a class of attractors, Trans. Amer. Math. Soc., 185 (1973), 183-198.

[28] Mallet-Paret, J., Negatively invariant sets of compact maps and an extension of a theorem of Cartwright, J. Differential Eqns., 22 (1976), 331-348.

[29] Mallet-Paret, J., Generic and qualitative properties of retarded functional differential equations, in Symposium of Functional Differential Equations São Carlos, Aug. 1975, Colecão Atas, Sociedade Brasileira de Matematica, 1977.

[30] Mallet-Paret, J., Generic periodic solutions of functional differential equations, J. Differential Eqns., 25 (1977), 163-183.

[31] Mañé, R., On the dimension of the compact invariant sets of certain nonlinear maps, in Lecture Notes in Math., vol. 898, 230-242, Springer-Verlag, 1981.

[32] Massatt, P., Stability and fixed points of dissipative systems. J. Differential Eqns. 40 (1981), 217-231.

[33] Massatt, P., Attractivity properties of α-contractions. J. Differential Eqns. To appear.

[34] Massatt, P., Asymptotic behavior of a strongly damped nonlinear wave equation. J. Differential Eqns. To appear.

[35] Matano, H., Convergence of solutions of one-dimensional semilinear parabolic equations, J. Math. Kyoto Univ. 18 (1978), 221-227.

[36] Nussbaum, R. D., Periodic solutions of analytic functional differential equations are analytic, Michigan Math. J., 20 (1973), 249-255.

[37] Oliva, W. M., Functional differential equations on compact manifolds and an approximation theorem, J. Differential Eqns., 5 (1969), 483-496.

[38] Oliva, W. M., Functional differential equations - generic theory, in Dynamical Systems - An International Symposium, vol. I, 195-208, eds. L. Cesari, J. K. Hale and J. P. LaSalle, Academic Press, New York, 1976.

[39] Oliva, W. M., Retarded equations on the sphere induced by linear equations. Preprint, 1982.

[40] Oliva, W. M., The behavior at the infinity and the set of global solutions of retarded functional differential equations, in Symposium of Functional Differential Equations, 103-126, São Carlos, Aug. 1975, Coleção Atas, Sociedade Brasileira de Matematica, 1977.

[41] Oliva, W. M., Functional differential equations on manifolds, Atas da Sociedade Brasileira de Matematica, 1 (1971), 103-116.

[42] Oliva, W. M., Some open questions in the geometric theory of retarded functional differential equations. Proc. 10th Brazilian Colloq. Math., Poços de Caldas, July 1975.

[43] Oliva, W. M., Stability of Morse-Smale maps. Preprint 1982.

[44] Palis, J., On Morse-Smale dynamical systems, Topology 8 (1969), 385-405.

[45] Palis, J. and S. Smale, Structural stability theorems in Global Analysis, Proc. Symp. in Pure Math. 14 (1970) Amer. Math. Soc., Providence, R.I.

[46] Palis, J. and F. Takens, Stability of parametrized families of gradient vector fields, Annals Math. To appear.

[47] Peixoto, M. M., On an approximation theorem of Kupka and Smale, J. Differential Eqns., 3 (1966), 214-227.

[48] Popov, V. M., Pointwise degeneracy of linear, time invariant, delay differential equations. J. Differential Eqns. 11 (1972), 541-561.

[49] Smale, S., An infinite dimensional version of Sard's Theorem, Amer. J. Math., 87 (1965), 861-866.

[50] Yorke, J., Noncontinuable solutions of differential-delay equations, Proc. Amer. Math. Soc., 21 (1969), 648-657.

[51] Zelenyak, T. I., Stabilization of solutions of boundary value problems for a second order parabolic equation with one space variable. Differential Equations 4 (1968), 17-22 (translated from Differentialniya Uravneniya).

[52] Chernoff, P. R. and J. E. Marsden, Properties of Infinite Dimensional Hamiltonian Systems. Lecture Notes in Math.,Vol. 425, Springer-Verlag, Berlin, 1974.

[53] Hale, J. K. and J. Scheurle, Smoothness of bounded solutions of nonlinear evolution equations, LCDS Report #83-12, in preparation, to be submitted to J. Differential Equations.

Appendix

An Introduction to the Homotopy Index Theory in Noncompact Spaces

Krzysztof P. Rybakowski

This appendix serves to introduce the reader to the main aspects of the homotopy index theory.

In its original form for (two-sided) flows on compact or locally compact spaces the theory is due mainly to Conley, although people like R. Easton, R. Churchill, J. Montgomery and H. Kurland should also be mentioned. The interested reader is referred to the monograph [Co] for an account of the original version of the theory.

Conley's theory, in its original form, was developed primarily for ODEs. By means of some special constructions, certain parabolic PDEs and RFDEs can also be treated in this original version of the theory. However, this imposes severe restrictions on the equations like, for example, the existence and knowledge of a bounded positively invariant set.

In papers [R1] - [R7], [RZ], Conley's theory was extended to large classes of semiflows on noncompact spaces. In particular, not only RFDEs and parabolic PDEs, but also certain classes of NFDEs and hyperbolic equations can be treated quite naturally by this extended theory. In the above cited papers, some applications to all these classes of equations are given.

We may consider Conley's original version of the homotopy index to be a generalization of the classical Morse index theory on compact manifolds: Morse assigns an index to every nondegenerate equilibrium of a gradient system, Conley assigns an index to every compact isolated invariant set of a not necessarily gradient ODE.

The extended homotopy index theory is, in a sense, analogous to the Palais-Smale extension of the classical Morse index to noncompact spaces.

Although our only application will be to RFDEs on $\mathbb{R}^m$, we will present the theory for general semiflows. This will clarify the main ideas. We begin with a well-known concept:

Definition 1. Given a pair (X,π) π is called a local semiflow (on X) if the following properties hold:

1. X is a topological space, $\pi: D \to X$ is a continuous mapping, D being an open subset of $\mathbb{R}^+ \times X$. (We write $x\pi t$ for $\pi(t,x)$.)
2. For every $x \in X$ there is an ω_x, $0 < \omega_x \leq \infty$, such that $(t,x) \in D$ if and only if $0 \leq t < \omega_x$.
3. $x\pi 0 = x$ for $x \in X$.
4. If $(t,x) \in D$ and $(s,x\pi t) \in D$, then $(t+s,x) \in D$ and $x\pi(t+s) = (x\pi t)\pi s$.

Remark. If $\omega_x = \infty$ for all $x \in X$, then π is called a (global) semiflow (on X).

(Local) semiflows are also called (local) dynamical (or, more appropriately, (local) semidynamical) systems.

Example 1. Let F be a locally Lipschitzian RFDE on an m-dimensional manifold M and let Φ be the corresponding solution map. Write $\varphi\pi_F t = \Phi_t\varphi$, whenever the right-hand side is defined. Then π_F is a local semiflow on C^0 (cf. Theorem 2.2 of these notes). We call π_F the local semiflow generated by the solutions of F. We omit the subscript $_F$ and write $\pi = \pi_F$ if no confusion can arise. If $M = \mathbb{R}^m$, then $F(\varphi) = (\varphi(0),f(\varphi))$, where $f: C^0 \to \mathbb{R}^m$ is locally Lipschitzian. In this case, we will write π_f instead of π_F.

In previous sections of these Notes, several important concepts were defined relative to the local semiflow π_F, like that of a solution and of an invariant set. It is useful to extend these concepts to general local semiflows π on a topological space X. In particular, let $\mathcal{J}$ be an interval in $\mathbb{R}$ and $\sigma: \mathcal{J} \to X$ be a mapping. σ is called a solution (of π) if for all $t \in \mathcal{J}$, $s \in \mathbb{R}^+$ for which $t+s \in \mathcal{J}$, it follows that $\sigma(t)\pi s$ is defined and $\sigma(t)\pi s = \sigma(t+s)$. If $0 \in \mathcal{J}$ and $\sigma(0) = x$ then we may say that σ is a solution through x. If $\mathcal{J} = (-\infty,\infty)$, then σ is called a global (or full) solution.

If Y is a subset of X, then set:

$$I^+(Y) = \{x \in X \mid x\pi[0,\omega_x) \subset Y\}$$

$$I^-(Y) = \{x \in X \mid \text{there is a solution } \sigma: (-\infty,0] \to X \text{ through } x \text{ with } \sigma(-\infty,0] \subset Y\}.$$

$$I(Y) = I^+(Y) \cap I^-(Y).$$

Y is called positively invariant if $Y = I^+(Y)$,

Y is called negatively invariant if $Y = I^-(Y)$,

Y is called invariant if $Y = I(Y)$.

In particular, if $\omega_x = \infty$ for every $x \in Y$, then Y is invariant iff for every $x \in Y$ there exists a full solution σ through x for which $\sigma(\mathbb{R}) \subset Y$.

For a general subset Y of X, $I^+(Y)$ (resp. $I^-(Y)$, resp. $I(Y)$) is easily seen to be the largest positively invariant (resp., negatively invariant, resp. invariant subset of Y). $I^+(Y)$ (resp. $I^-(Y)$) is often called the stable (resp. unstable) manifold of $K = I(Y)$, relative to Y.

To illustrate these concepts with an example, suppose that f is an RFDE on $\mathbb{R}^m$ of class C^1 and 0 is a hyperbolic equilibrium of f (cf.

[H1], Chapter 10). Then the well-known saddle-point property implies that there is a direct sum decomposition $C^0 = U \oplus S$ and a closed neighborhood Y of 0 such that $K = \{0\}$ is the largest invariant set in Y, i.e., $\{0\} = I(Y)$. Moreover, the sets $I^+(Y)$ and $I^-(Y)$ are tangent to S, resp. to U, at zero. There is a small ball $B_\delta \subset Y$ such that $I^+(Y) \cap B_\delta$ (resp. $I^-(Y) \cap B_\delta$) are diffeomorphic to $S \cap B_\delta$ (resp. $U \cap B_\delta$). Finally, the ω-limit set of every solution starting in $I^+(Y)$ (resp. the α-limit set of every solution defined on $(-\infty,0]$ and remaining in $I^-(Y)$) is equal to $\{0\}$. Therefore, the qualitative picture near the equilibrium looks as in Fig. 1.

The set $K = \{0\}$ has the important property of being isolated by Y. More generally, if K is a closed invariant set and there is a neighborhood U of K such that K is the largest invariant set in U, then K is called an <u>isolated invariant set</u>. On the other hand, if N is a closed subset of X and N is a neighborhood of $K = I(N)$, i.e., if the largest invariant set in N is actually contained in the interior of N, then N is called an <u>isolating neighborhood of</u> K. Hence, in the situation of Fig. 1, $K = \{0\}$ is an isolated invariant set and Y is an isolating neighborhood of K. Let us analyze the example a little further: The isolating neighborhood Y is rather arbitrary, i.e., its boundary ∂Y is unrelated in any way to the semiflow π. However, Fig. 1 suggests that one should be able to choose the set Y in such a way that ∂Y is "transversal" to π, i.e., such that orbits of solutions of π cross Y in one or the other direction (Fig. 2). In fact, this is, for example, the case for ODEs, where such special sets, called isolating blocks are used in connection with the famous Ważewski principle. The transversality of ∂Y with respect to π

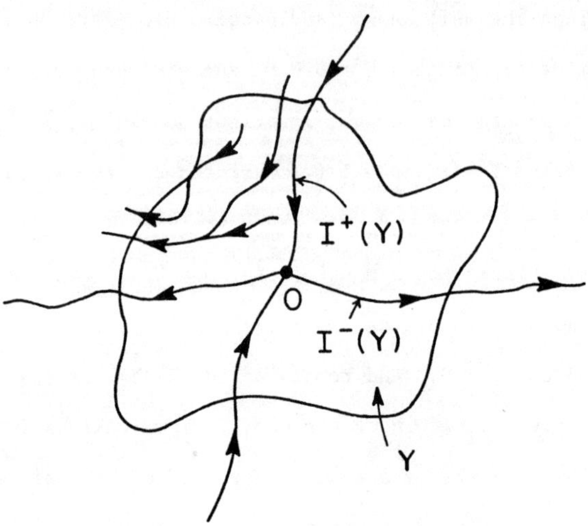

Figure 1

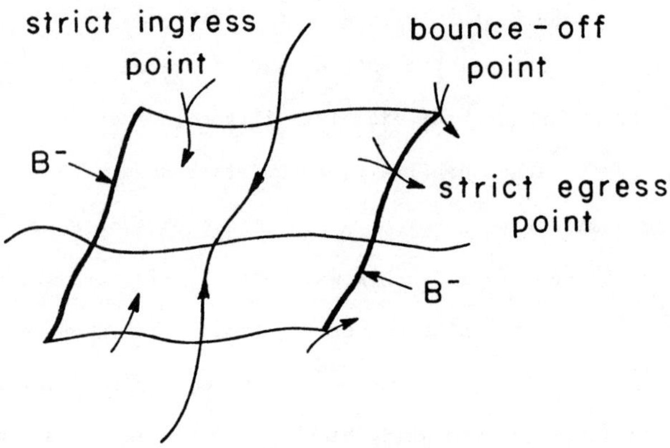

Figure 2

implies that every point x is of one of the following three types: it is either a strict egress, or a strict ingress or a bounce-off point.

Let us define those three concepts for an arbitrary local semiflow π.

Let $B \subset X$ be a closed set and $x \in \partial B$ a boundary point. Then x is called a strict egress (resp. strict ingress, resp. bounce-off) point of B, if for every solution $\sigma: [-\delta_1, \delta_2] \to X$ through $x = \sigma(0)$, with $\delta_1 \geq 0$ and $\delta_2 > 0$ there are $0 \leq \varepsilon_1 \leq \delta_1$ and $0 < \varepsilon_2 \leq \delta_2$ such that for $0 < t \leq \varepsilon_2$:

$$\sigma(t) \notin B \quad (\text{resp. } \sigma(t) \in \text{int}(B), \text{ resp. } \sigma(t) \notin B),$$

and for $-\varepsilon_1 \leq t < 0$:

$$\sigma(t) \in \text{int}(B) \quad (\text{resp. } \sigma(t) \notin B, \text{ resp. } \sigma(t) \notin B).$$

By B^e (resp. B^i, resp. B^b) we denote the set of all strict egress (resp. strict ingress, resp. bounce-off) points of the closed set B. We finally set $B^+ = B^i \cup B^b$ and $B^- = B^e \cup B^b$.

We then have the following:

Definition 2 (Isolating block). A closed set $B \subset X$ is called an isolating block, if

(i) $\partial B = B^e \cup B^i \cup B^b$

(ii) B^e and B^i are open in ∂B.

Note that for general semiflows, $B^e \cap B^b$ may be nonempty, and consist of points $x \in B^e$ for which there is no solution defined for some negative times.

If B is an isolating block such that B^- is not a strong deformation retract of B, then there is a nonempty, positively invariant set in B.

This is an important special case of Ważewski principle and was one of the motivations for developing the homotopy index theory for ODEs. (Cf. [Co]).

Since ω-limit sets of compact trajectories are invariant sets, Ważewski principle plus some compactness assumptions imply that $I(B) \neq \emptyset$. Moreover, it is obvious that B is an isolating neighborhood of $I(B)$. The important converse problem arises: Given an isolated invariant set K, is there an isolating neighborhood set B of K which is an isolating block? Fig. 2 suggests that this should be the case for hyperbolic equilibria, but we will try to give a general answer.

For two-sided flows on compact manifolds, the existence of isolating blocks was first proved by Conley and Easton [CE]. The proof uses the theory of fibre bundles and it needs both the two-sidedness of the flow as well as the compactness of the underlying space in a very crucial way, i.e., it applies essentially only to ordinary differential equations in finite dimensions.

An alternative proof, still for ODEs, was given by Wilson and Yorke [WY]. These authors construct two special Liapunov functions V_1 and V_2 and define $B = \{x \mid V_1(x) \leq \varepsilon, V_2(x) \leq \varepsilon\}$, for some $\varepsilon > 0$. This resembles Ważewski's original idea to use isolating blocks in the form of the so-called regular polyfacial sets, i.e., sets whose boundaries consist, piecewise, of level surfaces of special Liapunov-like functions.

Although Wilson and Yorke still use compactness and the two-sidedness of the flow in an essential way, a portion of their proof can be utilized in generalizing the existence result for isolating blocks to semiflows on non-necessarily compact spaces.

The following elementary observation gives a first hint of how to proceed:

Proposition 1. Let π be a local semiflow on the metric space X, K be an isolated invariant set and N be an isolating neighborhood of K.

Suppose that there exist continuous functions $V_i: N \to \mathbb{R}$, $i = 1,2$, satisfying the following properties:

(i) If $\sigma: \mathcal{J} \to N$ is a solution (of π) and $V_1(\sigma(t)) \neq 0$ (resp. $V_2(\sigma(t)) \neq 0$) for all $t \in \mathcal{J}$, then $t \to V_1(\sigma(t))$ is strictly increasing (resp. $t \to V_2(\sigma(t))$ is strictly decreasing).

(ii) If $x \in N$, then $x \in K$ if and only if $V_1(x) = 0$ and $V_2(x) = 0$.

(iii) If $\{x_n\} \subset N$ is a sequence such that $V_1(x_n) \to 0$ and $V_2(x_n) \to 0$ as $n \to \infty$, then $\{x_n\}$ contains a convergent subsequence.

Under these hypotheses, there is an $\varepsilon_0 > 0$ such that whenever $0 < \varepsilon_1, \varepsilon_2 \leq \varepsilon_0$, then the set $B = \mathrm{Cl}\{x \in N \mid V_1(x) < \varepsilon_1,\ V_2(x) < \varepsilon_2\}$ is an isolating block for K (i.e., such that B is also an isolating neighborhood of K). ("Cl" denotes closure.)

Remark: Property (i) means that V_1 and V_2 are Liapunov-like functions for the semiflow, one of them increasing and the other decreasing along solutions of π. Property (iii) looks very much like the Palais-Smale condition (cf. [PS] or [ChH]). (iii) is automatically satisfied if N is a compact metric space, and it will lead us to the concept of admissibility which will enable us to extend the homotopy index theory to noncompact spaces and (one-sided) semiflows.

Let us sketch the proof: first observe that there is an $\varepsilon_0 > 0$ such that

$$B_{\varepsilon_0} := \mathrm{Cl}\{x \in N \mid V_1(x) < \varepsilon_0 \text{ and } V_2(x) < \varepsilon_0\} \subset \mathrm{Int}\, N. \qquad (1)$$

In fact, if this is not true, then there exists a sequence $\{x_n\} \subset N$ such that $V_1(x_n) \to 0$ and $V_2(x_n) \to 0$ as $n \to \infty$, but $x_n \in \partial N$ for all n. Hence, by property (iii), we may assume that $\{x_n\}$ converges to some $x \in N$. By continuity, $V_1(x) = 0 = V_2(x)$. Hence, by (ii), $x \in K$. However, $x \in \partial N$ which is a contradiction (since N isolates K and therefore $\partial N \cap K = \emptyset$). Let $0 < \varepsilon_1, \varepsilon_2 \leq \varepsilon_0$ be arbitrary and set

$$B = \mathrm{Cl}\{x \in N \mid V_1(x) < \varepsilon_1,\ V_2(x) < \varepsilon_2\}.$$

To prove that B is an isolating block for K, note first that $K \subset \mathrm{Int}\, B$ (by (ii)). Moreover, by (1),

$$\partial B = \{x \in B \mid V_1(x) = \varepsilon_1 \text{ or } V_2(x) = \varepsilon_2\}.$$

Now, using property (i), it is easily proved that

$$B^e = \{x \in \partial B \mid V_1(x) = \varepsilon_1 \text{ and } V_2(x) < \varepsilon_2\}$$

$$B^i = \{x \in \partial B \mid V_2(x) = \varepsilon_2 \text{ and } V_1(x) < \varepsilon_1\}$$

$$B^b \supset \{x \in \partial B \mid V_1(x) = \varepsilon_1 \text{ and } V_2(x) = \varepsilon_2\}.$$

This implies that B is an isolating block and completes the proof.

Using Proposition 1, let us now prove the existence of an isolating block in the simplest case of a hyperbolic equilibrium of a linear RFDE. This will illustrate some of the ideas of the general case without introducing any technicalities:

<u>Proposition 2.</u> <u>If</u> 0 <u>is a hyperbolic equilibrium of the linear RFDE</u>

$$\dot{x} = Lx_t \tag{2}$$

on $\mathbb{R}^n$, then there exist arbitrarily small isolating blocks for $K = \{0\}$.

Proof: Let π be the semiflow generated by (2). It is a global semiflow. By results in [H1] there is a direct sum decomposition $C^0 = U \oplus S$, $\dim U < \infty$, such that $\Phi(t)U \subset U$ and $\Phi(t)S \subset S$, for $t \geq 0$, $\Phi(t)|_U$ can be uniquely extended to a group of operators, and there are constants $M, \alpha > 0$ such that

$$\begin{aligned} ||\Phi(t)\varphi|| &\leq Me^{-\alpha t}||\varphi|| \quad \text{for } \varphi \in S,\ t \geq 0 \\ ||\Phi(t)\varphi|| &\leq Me^{+\alpha t}||\varphi|| \quad \text{for } \varphi \in U,\ t \leq 0. \end{aligned} \tag{3}$$

Let $k = \dim U$ and $\Psi: \mathbb{R}^k \to U$ be a linear isomorphism. If $A_1: U \to U$ is the infinitesimal generator of the group $\Phi(t)|_U$, $t \in \mathbb{R}$, then there exists a $k \times k$-matrix B such that $\Psi^{-1}A_1\Psi = B$. It follows that $\operatorname{re} \sigma(B) > 0$. Hence there exists a positive definite matrix D such that $B^TD + DB = I$, where I is the identity matrix.

Now choose $\tau > 0$ such that $M\cdot(t+1)e^{-\alpha t} < 1/2$ for $t \geq \tau$. For $\varphi \in C^0$ define

$$V_1(\varphi) = (\Psi^{-1}P_U\varphi)^T D(\Psi^{-1}P_U\varphi), \tag{4}$$

$$V_2(\varphi) = \sup_{0\leq t\leq\tau} [(t+1)||\Phi(t)P_S\varphi||], \tag{5}$$

where P_U and P_S are the projections onto U, S resp., corresponding to the above direct sum decomposition. An easy computation shows that

$$\liminf_{t\to 0^+} \frac{1}{t}(V_1(\Phi(t)\varphi) - V_1(\varphi)) > 0 \quad \text{if} \quad \varphi \notin S$$

and

$$\limsup_{t\to 0^+} \frac{1}{t}(V_2(\Phi(t)\varphi) - V_2(\varphi)) < 0 \quad \text{if} \quad \varphi \notin U$$

Therefore V_1 and V_2 are easily seen to satisfy all assumptions of Proposition 1 (property (iii) follows from the fact that U is finite-dimensional). The proposition is proved.

If we try to prove the existence of isolating blocks for general semiflows by using Proposition 1, we have to find an hypothesis which implies property (iii) of that Proposition. Such an hypothesis can be formulated by means of the following fundamental concept:

Definition 3. Let π be a local semiflow on the metric space X, and N be a closed subset of X. N is called π-admissible (or simply admissible, if no confusion can arise) if for every sequence $\{x_n\} \subset X$ and every sequence $\{t_n\} \subset \mathbb{R}^+$ the following property is satisfied: if $t_n \to \infty$ as $n \to \infty$, $x_n \pi t_n$ is defined and $x_n \pi [0,t_n] \subset N$ for all n, then the sequence of end-points $\{x_n \pi t_n\}$ has a convergent subsequence.

N is called strongly π-admissible (or strongly admissible) if N is π-admissible and π does not explode in N, i.e., if whenever $x \in N$ is such that $\omega_x < \infty$, then $x\pi t \notin N$ for some $t < \omega_x$.

Remark. Admissibility is an asymptotic compactness hypothesis: if the solution through x_n stays in N long enough ($t_n \to \infty$!) then $\{x_n \pi t_n\}$ is a relatively compact set. Obviously, every compact set N in X is admissible, hence the concept is trivial for ODEs. However, bounded sets N are π-admissible for many semiflows in infinite dimensions, like the semiflows generated by RFDEs as in Example 1 (see below) or those generated by certain neutral equations and many classes of parabolic and even hyperbolic PDEs. (See [R1], [R4], [R5], [H2].)

The assumption that π does not explode in N is quite natural and it implies that as long as we stay in N, we can treat π like a global semiflow. However, it is useful for the applications not to assume a priori that π is a global semiflow on X, since many local semiflows cannot be modified outside a given set N without destroying their character (e.g., the fact that they are generated by a specific equation).

Example 1 (cont.). Suppose that $M = \mathbb{R}^m$ and $f: C^0 \to \mathbb{R}^m$ is locally Lipschitzian. Let $N \subset C^0$ be closed and bounded and $f(N)$ be bounded. Then the Arzelà-Ascoli Theorem easily implies that N is π_f-admissible (cf. the proof of Theorem 3.6.1 in [H1]). Moreover, Theorem 2.3.2 in [H1] (or rather its proof) implies that π_f does not explode in N. It follows that N is strongly π_F-admissible.

Similar statements are of course true for general manifolds M. They are related to the fact that under quite natural assumptions, the solution operators $\Phi(t)$, $t \geq r$ are conditionally compact (cf. Theorem 2.3 of these Notes).

More generally, if π is a (local) semiflow on a complete metric space X and its solution operator $T(t_0)$ is, for some $t_0 > 0$, a conditional α-contraction, then every bounded set $N \subset X$ is π-admissible (see Section 5 of these notes).

Let us note the following simple

Lemma 1. ([R1]). *If $N \subset X$ is closed and strongly admissible, then $I^-(N)$ and $I(N)$ are compact.*

In other words, the largest invariant set $K = I(N)$ in N and its unstable manifold relative to N are both compact.

Proof: If $\{y_n\} \subset I^-(N)$, then there are solutions $\sigma_n\colon (-\infty,0] \to N$ such that $\sigma_n(0) = y_n$, $n \geq 1$. Let $x_n = \sigma_n(-n)$. Obviously $x_n\pi[0,t_n] \subset N$ and $t_n \to \infty$, where $t_n = n$. Hence admissibility implies that $x_n\pi t_n = y_n$, $n \geq 1$, is a relatively compact sequence. Since the diagonalization procedure easily implies that both $I^-(N)$ and $I(N)$ are closed, the result follows.

Let us also note that for N as in Lemma 1, $\omega_x = \infty$ whenever $x \in I^+(N)$.

We are now in a position to state a main result on the existence of isolating blocks:

Theorem 1. ([R1]). If K is an isolated invariant set and N is a strongly admissible isolating neighborhood of K, then there exists an isolating block B such that $K \subset B \subset N$.

Hence we assert the existence of arbitrarily small isolating blocks for K as long as K admits a strongly admissible, but otherwise arbitrary, isolating neighborhood.

The proof of Theorem 1, given in [R1], is rather technical, but we should at least try to indicate its main ideas.

Let U be an open set such that $K \subset U$ and $\mathrm{Cl}\ U \subset N$, e.g., $U = \mathrm{Int}\ N$. Replacing N by $\mathrm{Cl}\ U$, if necessary, we may assume without loss of generality that $N = \mathrm{Cl}\ U$. Also, for the sake of simplicity, assume that π is a global semiflow. Finally, we may assume that $K \neq \emptyset$, since otherwise $B = \emptyset$ is an isolating block for K. Define the following mappings:

$$s_N^+\colon N \to \mathbb{R}^+ \cup \{\infty\}, \quad s_N^+(x) = \sup\{t \mid x\pi[0,t] \subset N\},$$

$$t_U^+\colon U \to \mathbb{R}^+ \cup \{\infty\}, \quad t_U^+(x) = \sup\{t \mid x\pi[0,t] \subset U\},$$

$$F\colon X \to [0,1], \quad F(x) = \min\{1, \operatorname{dist}(x, I^-(N))\},$$

$$G\colon X \to [0,1], \quad G(x) = \operatorname{dist}(x,K)/(\operatorname{dist}(x,K) + \operatorname{dist}(x, X\setminus N));$$

$$g_U^+(x) := \inf\{(1+t)^{-1}G(x\pi t) \mid 0 \le t < t_U^+(x)\},$$

$$g_N^-(x) := \sup\{\alpha(t)F(x\pi t) \mid 0 \le t \le s_N^+(x), \text{ if } s_N^+(x) < \infty,$$
$$\text{and } 0 \le t < \infty, \text{ if } s_N^+(x) = \infty\},$$

g_U^+ is defined on U, g_N^- is defined on N, $\alpha\colon [0,\infty) \to [1,2)$ is a fixed monotone C^∞-diffeomorphism.

Then the following lemma holds (see [R1]):

Lemma 2. (i) s_N^+ *is upper-semicontinuous*, t_U^+ *is lower-semicontinuous*.
(ii) g_U^+ *is upper-semicontinuous, and* g_U^+ *is continuous in a neighborhood of* K. *Moreover, if* $g_U^+(x) \ne 0$; *then*

$$\dot{g}_U^+(x) = \liminf_{t\to 0^+} (1/t)(g_U^+(x\pi t) - g_U^+(x)) > 0.$$

If $g_U^+(x) = 0$, *then for every* $t \in \mathbb{R}^+$, $x\pi t \in U$ *and* $g_U^+(x\pi t) = 0$.
(iii) g_N^- *is upper-semicontinuous. If* $t_U^+(x) = s_N^+(x)$ *on* U, *then* g_N^- *is continuous on* U.

Moreover, if $g_N^-(x) \ne 0$ *then*

$$\dot{g}_N^-(x) = \limsup_{t\to 0^+} (1/t)(g_N^-(x\pi t) - g_N^-(x)) < 0.$$

If $g_N^-(x) = 0$, *then for every* $t \le s_N^+(x)$, $g_N^-(x\pi t) = 0$.

Therefore, taking N_1 to be an appropriate isolating neighborhood of K such that $N_1 \subset U$ and defining V_1 and V_2 to be the restrictions to N_1 of g_U^+ and g_N^- resp., we see that all assumptions of Proposition 1 are satisfied except that maybe V_2 is not continuous. In particular, property

(iii) of that proposition is a consequence of the fact that $I^-(N)$ is compact (see Lemma 1 above). Therefore, the set

$$\tilde{N} = \text{Cl}\,\tilde{U}, \text{ where } \tilde{U} = \{x \in N_1 \mid V_1(x) < \varepsilon,\ V_2(x) < \varepsilon\}, \quad \varepsilon \text{ small},$$

is not an isolating block, in general. However, this set has some properties of an isolating block, e.g., that $t^+_{\tilde{U}}(x) = s^+_{\tilde{N}}(x)$ on $\tilde{U}$. Therefore, we can repeat the same process by taking $\tilde{N}$ to be a new isolating neighborhood of K, and defining $g^+_{\tilde{U}}$, $g^-_{\tilde{N}}$ as above. Now Lemma 2, (iii) implies that $g^-_{\tilde{N}}$ is continuous on U. Hence taking $\tilde{N}_1$ to be an isolating neighborhood of K with $\tilde{N}_1 \subset \tilde{U}$ and letting $\tilde{V}_1$ and $\tilde{V}_2$ to be the restrictions to $\tilde{N}_1$ of $g^+_{\tilde{U}}$ and $g^-_{\tilde{N}}$, we can satisfy all the hypotheses of Proposition 2, thus proving the theorem.

If a set $K \neq \emptyset$ satisfies the assumptions of Theorem 1 then, of course, there are infinitely many isolating blocks for K. Moreover, if we perturb the semiflow π a little (for instance, by perturbing the right-hand side of an RFDE) then an isolating block with respect to the unperturbed semiflow, in general, is no longer an isolating block with respect to the perturbed semiflow. However, all isolating blocks for a given set K have a common property which may roughly be described as follows: take an admissible isolating block B for K, and collapse the subset B^- of B to one point. Then the resulting quotient space B/B^- is independent of the choice of B, modulo homeomorphisms or deformations preserving the base points [B]. Therefore the homotopy type of B/B^- is independent of the choice of B and this homotopy type is what we call the homotopy index of K.

Before giving a precise definition of the homotopy index, let us recall a few concepts from algebraic topology.

Definition 4. Let Y be a topological space and Z be a closed set in Y. If $Z \neq \emptyset$, then define Y/Z to be the set of all equivalence classes of the following equivalence relation $x \sim y$ iff $x = y$ or $x,y \in Z$. Y/Z is endowed with the quotient space topology. If $Z = \emptyset$, choose any point $p \notin Y$, give the union $Y \cup \{p\}$ the sum topology and set $Y/Z = Y/\emptyset := (Y \cup \{p\})/\{p\}$.

Let $[Z]$ denote either the equivalence class of Z in Y/Z (if $Z \neq \emptyset$), or else the equivalence class of $\{p\}$.

Then in each case the space Y/Z is regarded as a pointed space with the distinguished base point $[Z]$.

Definition 5. Let (X,x_0) and (Y,y_0) be two pointed spaces. We say that (X,x_0) and (Y,y_0) are homotopy equivalent if there exist continuous base point preserving maps $f: X \to Y$, $g: Y \to X$ such that $f \circ g$ and $g \circ f$ are homotopic (with base point preserving homotopies) to the respective identity maps. The homotopy type of (X,x_0), denoted by $h(X,x_0)$ is the class of all pointed spaces which are homotopy equivalent to (X,x_0).

Definition 6. Let X be a metric space. Then $\mathscr{S}$ (or $\mathscr{S}(X)$, if confusion may arise) is the set of all pairs (π,K), where π is a local semiflow on X, and K is an isolated, π-invariant set admitting a strongly π-admissible isolating neighborhood.

Remark. $\mathscr{S}$ is the class of all (π,K) satisfying the hypotheses of Theorem 1. Therefore, there exists a strongly admissible isolating block B for K.

Now we have the following uniqueness result:

Theorem 2. ([R1], [R2]). Let $(\pi,K) \in \mathscr{S}$, and B,, $\tilde{B}$ be two strongly admissible isolating blocks for K (relative to the semiflow π). Then $(B/B^-, [B^-])$ and $(\tilde{B}/\tilde{B}^-, [\tilde{B}^-])$ are homotopy equivalent. Consequently, the homotopy type $h(B/B^-, [B^-])$ only depends on the pair $(\pi,K) \in \mathcal{S}$ and we write

$$h(\pi,K) = h(B/B^-, [B^-])$$

$h(\pi,K)$ is called the homotopy index of (π,K).

Remark. If π is clear from the context, we write $h(K) = h(\pi,K)$ and speak of the homotopy index of K.

Before giving a few hints about the proof of Theorem 2, let us compute the index of $K = \{0\}$ in Proposition 2.

In fact, by Proposition 1, the set

$$B = \mathrm{Cl}\{\varphi \in C^0 \mid V_1(\varphi) < 1,\ V_2(\varphi) < 1\}$$

is a isolating block for $K = \{0\}$. But B is easily seen to be equal to $\{\varphi \in C^0 \mid V_1(\varphi) \leq 1,\ V_2(\varphi) \leq 1\}$. Moreover,

$$B^- = \{\varphi \in B \mid V_1(\varphi) = 1\}$$

Let $H\colon B \times [0,1] \to B$ be defined as $H(\varphi,s) = P_U\varphi + (1-s)P_S\varphi$.

Since $H(B^- \times [0,1]) \subset B^-$, H induces a continuous, base point preserving homotopy $\tilde{H}\colon B/B^- \times [0,1] \to B/B^-$. $\tilde{H}$ is a strong deformation retraction of $(B/B^-, [B^-])$ onto $(B_1/\partial B_1, [\partial B_1])$ where

$$B_1 = \{\varphi \in U \mid V_1(\varphi) \leq 1\}.$$

Now B_1 is an ellipsoid in $U \cong \mathbb{R}^k$, hence the pair $(B_1,\partial B_1)$ is homeomorphic to (E^k,S^{k-1}), where E^k is the unit ball, and S^{k-1} is the unit sphere in $\mathbb{R}^k$.

Now E^k/S^{k-1} is homeomorphic to the pointed k-dimensional sphere

(S^k, s_0) (with a base-point preserving homeomorphism). Altogether we obtain that

$$h(B/B^-, [B^-]) = h(B_1/\partial B_1, [\partial B_1])$$
$$= h(E^k/S^{k-1}, [S^{k-1}]) = h(S^k, s_0) =: \Sigma^k.$$

We obtain the following corollary:

Corollary 1. (cf. [R4]). Under the assumptions of Proposition 2

$$h(\pi, \{0\}) = \Sigma^k$$

where $k = \dim U$ and Σ^k is the homotopy type of a pointed k-sphere (S^k, s_0).

Hence the homotopy index of a hyperbolic equilibrium of a linear RFDE is determined by the dimension of its unstable manifold. This result is of crucial importance in the applications to be discussed later.

Let us now indicate a few ideas involved in the proof of Theorem 2.

Let B be a strongly admissible, isolating block for K. Write $N_1 = B$, $N_2 = B^-$. For $t \geq 0$, let N_2^{-t}, called the t-exit ramp of (N_1, N_2), be the set of all $x \in N_1$ such that $x\pi s \in N_2$ for some $0 \leq s \leq t$ (see Fig. 3). Moreover, let $N_1^t = \{x \in N_1 \mid$ there is a y such that $y\pi[0,t] \subset N_1$ and $y\pi t = x\}$. (See Fig. 4.) Figures 3 and 4 suggest that $N_1^t \setminus N_2^{-t}$ can be made arbitrarily small in the sense that whenever V is an arbitrary neighborhood of K, then $N_1^t \setminus N_2^{-t} \subset V$ for large t. This may be described as squeezing the block B. Of course, the result of the squeezing is not a block. However, by using the semiflow π as a natural homotopy mapping, we can prove that $B/B^- = N_1/N_2$ is homotopy equivalent to N_1^t/N_2^{-t}. If $\tilde{B}$ is another block, then, by what we said above, $N_1^t \setminus N_2^{-t} \subset \tilde{B}$ for large t. Therefore we

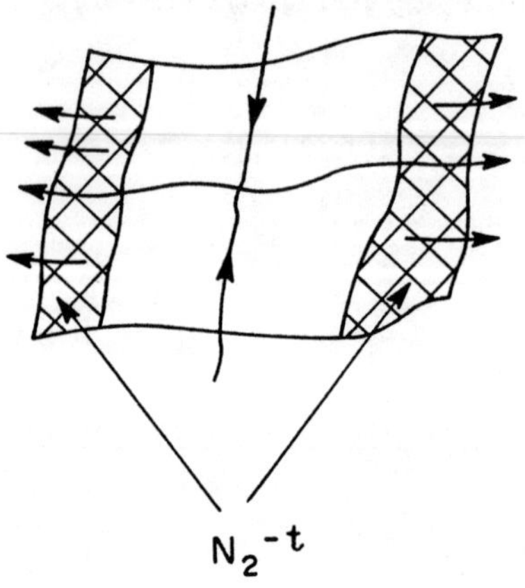

Figure 3

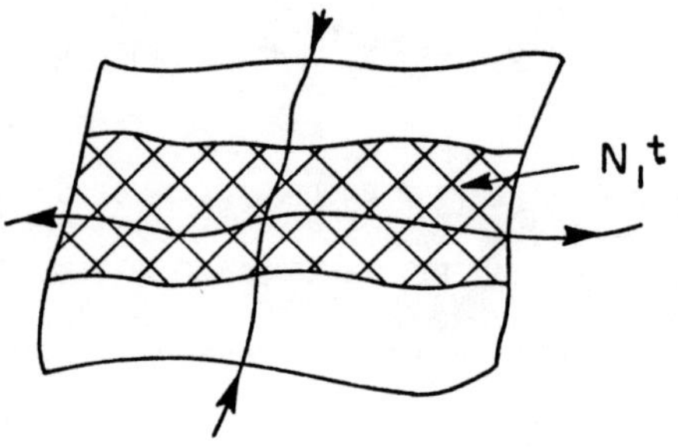

Figure 4

obtain a mapping $f: B/B^- \to \tilde{B}/\tilde{B}^-$ roughly as a composition of a "squeezing" followed by an inclusion. Similarly, a mapping $g: \tilde{B}/\tilde{B}^- \to B/B^-$ is defined. The deformation nature of the squeezing implies that $f \circ g$ and $g \circ f$ are homotopic to the corresponding identity maps. This proves Theorem 2.

Let us note that the pairs (N_1, N_2^{-t}) and (N_1^s, N_2^{-t}) generated by the isolating block B inherit certain properties of the pair (B, B^-). Such pairs are called index or quasi-index pairs (for K). Hence (B, B^-) is a special index pair for K. A precise definition of index and quasi-index pair is given in [R2]. One can show that whenever (N_1, N_2) is an index or quasi-index pair for K, then N_1/N_2 is homotopy equivalent to B/B^-, where B is an isolating block for B (of course, the usual admissibility assumption has to be imposed). Hence the homotopy index can be defined by general index or quasi-index pairs. However, the special index pairs (B, B^-) induced by isolating blocks have several advantages: e.g., they permit the use of arbitrary homology and cohomology modules, whereas only the Cech cohomology groups can be meaningfully used with general index pairs. This is due to the fact that the inclusion $B^- \subset B$ is a cofibration, a result which is not true for general index pairs.

The homotopy index as defined in Theorem 2 has an important property of being invariant under continuations of the semiflow. What is meant by this expression is that under certain admissible changes of the pair (π, K), the index $h(\pi, K)$ remains invariant. The changes of (π, K) are described by introducing a parameter $\lambda \mapsto (\pi(\lambda), K(\lambda))$ where λ varies over elements in some metric space Λ and $(\pi(\lambda), K(\lambda)) \in \mathcal{S}$. Call the resulting map α. When is α "admissible" in the sense that it leaves the index invariant?

A plausible condition is that 1) the map $\lambda \mapsto \pi(\lambda)$ is continuous in some sense (e.g., that $\pi(\lambda)$ represent RFDEs $F(\lambda)$ with continuously varying $\lambda \to F(\lambda)$), and 2) that there is a set N such that no bifurcations of invariant sets occur at the boundary of N as λ is varied.

This situation is analogous to that of the Leray-Schauder fixed point index, which remains constant under homotopies as long as no fixed-points appear on the boundary of the set considered. We need a third, technical assumption, which is, in a sense, a collective admissibility condition on N. More precisely, we have

Definition 7. Let X be a metric space and N be a closed set in X. Let $\{\pi_n\}$ be a sequence of local semiflows on X. N is called $\{\pi_n\}$-admissible if for every choice of sequences $\{x_n\} \subset X$, $\{t_n\} \subset \mathbb{R}^+$ satisfying (i) $t_n \to \infty$, as $n \to \infty$, (ii) $x_n \pi_n t_n$ is defined and $x_n \pi_n [0,t_n] \subset N$ for every n, it follows that the sequence $\{x_n \pi_n t_n\}$ of endpoints has a convergent subsequence.

Of course, if $\pi_n \equiv \pi$ for all n, then this definition reduces to the admissibility condition given previously.

Example 1 (cont.). Let $M = \mathbb{R}^m$ and $f_n : C^0 \to \mathbb{R}^m$, $n \geq 1$, be a sequence of locally Lipschitzian maps. If $N \subset C^0$ is a closed bounded set such that the set $\bigcup_{n=1}^{\infty} f_n(N)$ is bounded, then N is $\{\pi_n\}$-admissible where $\pi_n = \pi_{f_n}$.

This follows, as before in the case of one semiflow, by an application of the Arzela-Ascoli theorem.

We can now formulate

Definition 8. Let Λ be a metric space and $\alpha: \Lambda \to \mathcal{S}$ be a mapping. Since for $\lambda \in \Lambda$, $\alpha(\lambda) = (\pi(\lambda), K(\lambda))$, we write $\alpha_1(\lambda) = \pi(\lambda)$, $\alpha_2(\lambda) = K(\lambda)$. Let

$\lambda_0 \in \Lambda$. We say that α is $\mathcal{S}$-continuous at λ_0 if there is an isolating neighborhood N of $\alpha_2(\lambda_0)$ (relative to $\alpha_1(\lambda_0)$), and a neighborhood W of λ_0 in Λ such that the following properties hold:

1) For every $\lambda \in W$, N is an isolating neighborhood of $\alpha_2(\lambda)$, relative to $\alpha_1(\lambda)$, and N is strongly $\alpha_1(\lambda)$-admissible.

2) For every sequence $\{\lambda_n\} \subset W$ converging to λ_0:

(2.1) N is $\{\alpha_1(\lambda_n)\}$-admissible.

(2.2) the sequence $\{\alpha_1(\lambda_n)\}$ of local semiflows converges to the local semiflow $\alpha_1(\lambda_0)$, as $n \to \infty$.

We remark that if $\{\pi_n\}$ is a sequence of local semiflows on X, then we say that π_n <u>converges to the local semiflow</u> π <u>as</u> $n \to \infty$ ($\pi_n \to \pi$, <u>as</u> $n \to \infty$) if whenever $x_n \to x$ in X, $t_n \to t$ in $\mathbb{R}^+$ and $x\pi t$ is defined, then $x_n\pi_n t_n$ is defined for n sufficiently large, and $x_n\pi_n t_n \to x\pi t$.

This, in fact, is a very weak type of convergence, e.g., we have

Example 1. (cont.). Let $f_n\colon C^0 \to \mathbb{R}^m$, $n \geq 0$, be a sequence of locally Lipschitzian mappings such that $f_n(\varphi) \to f_0(\varphi)$, as $n \to \infty$, uniformly on compact subsets of C^0. Then $\pi_n \to \pi_0$ as $n \to \infty$, where $\pi_n = \pi_{f_n}$, $n \geq 0$. This is an easy exercise left to the reader.

Definition 8 gives precise conditions on the map α to be "admissible". In fact we have

Theorem 3 ([R1], [R2]). *If $\alpha\colon \Lambda \to \mathcal{S}$ is $\mathcal{S}$-continuous (i.e., if α is $\mathcal{S}$-continuous at λ_0, for every $\lambda_0 \in \Lambda$), then the index $h(\alpha(\lambda))$ is constant on connected components of Λ. In other words, if λ_1, λ_2 belong to the same connected component of Λ, then $h(\alpha(\lambda_1)) = h(\alpha(\lambda_2))$.* (Note $\alpha(\lambda) = (\pi(\lambda), K(\lambda))$).

In particular, if $\Lambda = [0,1]$ then $h(\alpha(0)) = h(\alpha(1))$. This latter relation is basic in the applications of the index. The idea is, of course, to "deform" (or "continue") a given equation to a simpler equation for which the index is known. This will yield the index with respect to the original system.

Not even an intuitive description of the proof of Theorem 3 can be given here.

Before turning to some applications of the index, let us state a result which shows that, in a certain sense, the homotopy index is a finite-dimensional concept:

Theorem 4. ([R1]). Let $(\pi, K) \in \mathcal{S}$ and B be a strongly admissible isolating block. Then the natural inclusion and projection mappings include the following isomorphisms of the Čech cohomology:

$$\begin{aligned} H^*(B/B^-, \{[B^-]\}) &\cong H^*(B, B^-) \\ &\cong H^*(B \cap I^-(B),\ B^- \cap I^-(B)) \\ &\cong H^*((B \cap I^-(B))/(B^- \cap I^-(B)),\ \{[B^- \cap I^-(B)]\}). \end{aligned}$$

Recall that $I^-(B)$ is the unstable manifold of K relative to B.

Theorem 4 is also valid for arbitrary index pairs (N_1, N_2). The proof follows by an application of the tautness and continuity properties of the Čech cohomology.

To see the significance of Theorem 4 suppose that X is an open subset of a Banach space E. Moreover, assume that the solution operator $T(t_0)$ of π, is, for some $t_0 > 0$, a C^1-map whose derivative can be decomposed as a sum of a contraction and a compact map. (This is the case for semiflows generated by many RFDEs and NFDEs, but also by semilinear parabolic and even some hyperbolic PDEs.) Then Theorem 6.1 of these notes implies that $I^-(B)$ has finite Hausdorff dimension. Consequently, under these hypotheses, the Čech cohomology of the homotopy index is that of a finite-dimensional space. In particular, only finitely many of the groups $H^{*q}(B/B^-,\{[B^-]\})$ are nontrivial. This latter result also gives a heuristic explanation of why Ważewski's principle is applicable to many infinite-dimensional problems despite the fact that, e.g. the infinite dimensional unit sphere is a strong deformation retract of the closed unit ball.

We will now give a few applications of the homotopy index to RFDEs on $M = \mathbb{R}^m$. In previous sections, the union $A(F)$ of all global bounded orbits of the RFDE(F) was studied. Conditions were given to assure that $A(F)$ is bounded (hence compact), connected and attracts all compact sets. In this case $A(F)$ is a maximal (hence isolated) compact invariant set which has an attractor nature. In the next few pages, we will exhibit a class of RFDEs on $M = \mathbb{R}^m$, for which the set $A(F)$ is bounded, i.e. it is a maximal compact invariant set, but $A(F)$ is not necessarily an attractor. The condition roughly is asymptotic linearity of f and "non-criticality" at infinity. We will also compute the index of $A(F)$ and make some statement about the structure of $A(F)$. In particular, $A(F)$ will, with one exception, have a nonempty unstable manifold. Furthermore, although $A(F)$ need not be connected (we give an example of that) it is irreducible (index-connected), i.e. $A(F)$ cannot be decomposed as a disjoint union of two sets with nonzero homotopy index. Note that we will write $A(f)$ for $A(F)$, where $F(\varphi) = (\varphi(0), f(\varphi))$.

We begin with the following result:

Theorem 5. Consider a sequence f_n, $n = 1,2,\ldots$ of continuous mappings from C^0 to $\mathbb{R}^m$ such that every f_n is locally Lipschitzian, and let $\pi_n = \pi_{f_n}$ be the corresponding sequence of local semiflows on C^0.

Assume the following hypotheses:

(H1) There is a closed set $G \subset C^0$ such that for all $a \geq 0$, $a \cdot G \subset G$, and there is a continuous mapping $L: C^0 \to \mathbb{R}^m$ which is locally Lipschitzian and such that $L(a\varphi) = aL(\varphi)$ for $a \geq 0$, $\varphi \in G$.

(H2) For every $K > 0$, there is an $M > 0$ such that, for all n and every $\varphi \in G$ for which $||\varphi|| \leq K$, it follows that $||f_n(\varphi)|| \leq M$ and $||L(\varphi)|| \leq M$.

(H3) If $\varphi_n \in G$ and $||\varphi_n|| \to \infty$ as $n \to \infty$ then

$$\frac{||f_n(\varphi_n) - L(\varphi_n)||}{||\varphi_n||} \to 0 \quad \text{as} \quad n \to \infty.$$

(H4) If $t \to \sigma(t)$ is a bounded solution of π_L on $(-\infty,\infty)$, then $\sigma(t) \equiv 0$ for all $t \in R$.

Under these hypotheses, there is an $M_0 > 0$ and an n_0 such that, for all $n \geq n_0$, and every global bounded solution $t \to \sigma(t)$ of π_n such that $\sigma[\mathbb{R}] \subset G$, it follows that $\sup_{t \in \mathbb{R}} ||\sigma(t)|| \leq M_0$.

Remark. In the applications of Theorem 5 in this section, $G = C^0$ and L is a linear mapping, hence (H1) is automatically satisfied. However, we give the more general version of Theorem 5 with the view of possible applications to "nonnegative" mappings L. In these cases, G would be the "nonnegative cone" of C^0. In [R4], this general version has been applied to nonnegative solutions of parabolic PDEs.

Notice that in the statement of Theorem 5 as well as in its proof we use "$|| \; ||$" to denote both the euclidean norm in $\mathbb{R}^m$ and the induced sup-norm in C^0. Confusion should not arise.

Proof of Theorem 5: Notice first that if $t \to \sigma(t)$ is a global bounded solution of the semiflow π_f, then there is a unique continuous mapping $x: (-\infty,\infty) \to \mathbb{R}^n$ such that $\sigma(t) = x_t$ for every $t \in \mathbb{R}$. Moreover $t \to x(t)$ is a solution of the RFDE (f) on $(-\infty,\infty)$. Now suppose that the theorem is not true. Then, taking subsequences if necessary, we may assume that there is a sequence of global bounded solutions $t \to \sigma_n(t) \in G$ of π_n such that $\alpha_n = \sup_{t\in\mathbb{R}} ||\sigma_n(t)|| \to \infty$, $\alpha_n \neq 0$, and $||\sigma_n(0)|| > \alpha_n - 1$. Let $x^n: (-\infty,\infty) \to \mathbb{R}^n$ be the corresponding sequence of mappings such that $x_t^n = \sigma_n(t)$ for $t \in$ Let $y^n(t) = \dfrac{x^n(t)}{\alpha_n}$, and $\tilde{\sigma}_n(t) = y_t^n$, $t \in \mathbb{R}$. By (H1), $\tilde{\sigma}_n(t) \in G$ for all n, and all $t \in \mathbb{R}$. Let $\tilde{f}_n(\varphi) = \dfrac{f_n(\alpha_n\varphi)}{\alpha_n}$. Then $\tilde{f}_n: C^0 \to \mathbb{R}^n$ is locally Lipschitzian. We will show that for every $\rho \geq 0$, $\sup_{\substack{\varphi\in G \\ ||\varphi||\leq\rho}} ||f_n(\varphi) - L(\varphi)|| \to 0$ as $n \to \infty$.

In fact, let $\varepsilon > 0$ be arbitrary. Then by hypothesis (H3) there are $K > 0$ and n_1 such that whenever $\varphi \in G$, $||\varphi|| > K$, $n \geq n_1$, then $(||f_n(\varphi) - L(\varphi)||)/||\varphi|| < \varepsilon/\rho$.

Moreover, by hypothesis (H2), there is an $M > 0$ such that for all n and all $\varphi \in G$, $||\varphi|| \leq K$: $||f_n(\varphi)|| \leq M$ and $||L(\varphi)|| \leq M$. Choose $n_2 \geq n_1$ such that $(2M)/\alpha_n < \varepsilon$ for $n \geq n_2$. Then we have for every $\varphi \in G$, $||\varphi|| \leq \rho$ and every $n \geq n_2$ (using hypothesis (H1):

$$||\tilde{f}_n(\varphi) - L(\varphi)|| = \alpha_n^{-1}\cdot(||f_n(\alpha_n\cdot\varphi) - L(\alpha_n\cdot\varphi)||) \leq (2M)/\alpha_n < \varepsilon,$$

if $\alpha_n\cdot||\varphi|| \leq K$,

$$||\tilde{f}_n(\varphi) - L(\varphi)|| = ||\varphi||\cdot(\alpha_n\cdot||\varphi||)^{-1}\cdot||f_n(\alpha_n\cdot\varphi)$$
$$-L(\alpha_n\cdot\varphi)|| < \rho\cdot\varepsilon/\rho = \varepsilon$$

if $\alpha_n\cdot||\varphi|| > K$. Hence our claim is proved.

Let $\tilde{\pi}_n = \pi_{\tilde{f}_n}$. Then it follows that $t \to \tilde{\sigma}_n(t)$ is a global bounded solution of $\tilde{\pi}_n$.

It follows from what we have just proved and from hypothesis (H2) (using Example 1 above) that N is $\{\pi_n\}$-admissible, for every closed bounded set $N \subset C^0$. Since $||\tilde{\sigma}_n(t)|| \leq 1$ for every $t \in \mathbb{R}$ and every n, it follows that for every $t \in \mathbb{R}$, the sequence $\{\tilde{\sigma}_n(t)\}$ is precompact in C^0. Hence $\{y^n\}$ is precompact on $[-r+t,t]$.

If $r > 0$, this means that $\{y^n\}$ is equicontinuous at every $t \in \mathbb{R}$, hence, using an obvious diagonalization procedure, we conclude that there exists a subsequence $\{y^{n_k}\}$ of $\{y^n\}$ and a continuous map $y\colon (-\infty,\infty) \to \mathbb{R}^m$ such that $y^{n_k}(t) \to y(t)$ as $k \to \infty$, uniformly on compact intervals.

If $r = 0$, then the RFDEs involved are in fact, ODEs and we obtain

$$||y^n(t) - y^n(t_0)|| \leq \int_{t_0}^{t} ||\tilde{f}_n(y_n(s))||ds.$$

Hence, again $\{y^n\}$ is equicontinuous at every $t \in \mathbb{R}$ and we obtain a y and a subsequence $\{y^{n_k}\}$ as above.

A simple limit argument now shows that $t \to y(t)$ is a global bounded solution of the RFDE (L), which in view of hypothesis (H4), implies that

$y(t) \equiv 0$. However,

$$||\tilde{\sigma}_{n_k}(0)|| = \alpha_{n_k}^{-1} \cdot ||\sigma(0)|| > \alpha_{n_k}^{-1}(\alpha_{n_k} - 1) \to 1 \quad \text{as} \quad k \to \infty.$$

Hence $||y_0^{n_k}|| \to 1$ as $k \to \infty$, a contradiction which proves the theorem.

Using results in [H1] we see that if $G = C^0$ and L is linear and bounded, then hypothesis (H4) is equivalent to the requirement that $\varphi = 0$ be a hyperbolic equilibrium of $\dot{x} = Lx_t$.

We thus have the following

Theorem 6. *Let* $f: C^0 \to \mathbb{R}^m$ *be a locally Lipschitzian and completely continuous mapping*. *Furthermore, let* $L: C \to \mathbb{R}^m$ *be a bounded linear mapping*. *Suppose that*

$$\lim_{||\varphi|| \to \infty} \frac{f(\varphi) - L(\varphi)}{||\varphi||} = 0$$

If zero is a hyperbolic equilibrium of L *and* d *is the dimension of the unstable manifold* U *of* L, *then* $A(f)$ *is bounded, hence compact*. *Moreover*, $h(\pi_f, A(f)) = \Sigma^d$.

Proof: Let $f_\sigma = (1-\sigma)f + \sigma L$, $\sigma \in [0,1]$, and let $\pi_\sigma = \pi_{f_\sigma}$. Write $\mathcal{J}_\sigma := A(f_\sigma)$. Using Theorem 5 and a simple compactness argument, it is easily seen that there is a closed bounded set $N \subset C^0$ such that $\mathcal{J}_\sigma \subset \text{Int } N$ and N is strongly π_σ-admissible for every $\sigma \in [0,1]$. Hence $(\pi_\sigma, \mathcal{J}_\sigma) \in \mathcal{S}$ and it is easily seen that the mapping $\alpha: \sigma \to (\pi_\sigma, \mathcal{J}_\sigma)$ is $\mathcal{S}$-continuous. It follows, by Theorem 3, that $h(\pi_0, \mathcal{J}_0) = h(\pi_1, \mathcal{J}_1)$. However, $(\pi_0, \mathcal{J}_0) = (\pi_f, A(f))$, $(\pi_1, \mathcal{J}_1) = (\pi_L, \{0\})$. Hence, by Corollary 1

$$h(\pi_f, A(f)) = \Sigma^d$$

and the proof is complete.

We will now draw a few conclusions from Theorem 6.

First, let us define the following concept.

Definition 9. A pair $(\pi, K) \in \mathcal{S}$ is called irreducible, if K cannot be decomposed as a disjoint union $K = K_1 \cup K_2$ of two compact sets (both these sets would necessarily be invariant) such that

$$h(\pi, K_1) \neq \bar{0} \quad \text{and} \quad h(\pi, K_2) \neq \bar{0}.$$

Let us remark that $\bar{0}$ is the homotopy type of a one-point pointed space. It is clear that e.g., $h(\pi, \emptyset) = \bar{0}$.

Definition 9 generalizes the concept of connectedness; in fact, if K is connected, then (π, K) is irreducible, of course.

Moreover, we have the following:

Proposition 3 (see [R5]). If $(\pi, K) \in \mathcal{S}$ and $h(\pi, K) = \bar{0}$ or $h(\pi, K) = \Sigma^k$ for some $k \geq 0$, then (π, K) is irreducible.

The purely algebraic-topological proof of Proposition 3 is omitted.

As a consequence of Proposition 3, we see that $(\pi_f, A(f))$ is irreducible. Later on we will see that $A(f)$ does not have to be connected. Still irreducibility implies the following

Proposition 4 (see [R5]). Assume that all hypotheses of Theorem 6 are satisfied. Let $K \subset A(f)$ be an isolated π-invariant set and suppose that

$$h(\pi_f, K) \neq \bar{0} \quad \text{and} \quad h(\pi_f, K) \neq \Sigma^d.$$

Then there exists a global bounded solution $t \to x(t)$ of the RFDE(f) such that for some t_0, $x_{t_0} \notin K$ but either the α- or the ω-limit set of $t \to x_t$ is contained in K (or maybe both).

In other words, although the orbit of $t \to x_t$ is not fully contained in K, it either emanates from K or tends to K, or both. If π_f is gradient-like, this means that there is a heteroclinic orbit joining a set of equilibria $L_1 \subset K$ with some other set of equilibria. In the special case that $K = \{0\}$, this also gives us existence of nontrivial equilibria of π_f. Incidentally, this procedure, applied to semilinear parabolic equations, proves the existence of nontrivial solutions of elliptic equations ([R4], [R7]).

The proof of Proposition 4 is obtained by noticing that if the proposition is not true, then, there exists a compact set K' disjoint from K and such that $K \cup K' = A(f)$. However, the irreducibility of $(\pi_f, A(f))$ then leads to a contradiction, since $h(\pi_f, K) \neq \overline{0}$ and $h(\pi_f, K') \neq \overline{0}$.

Proposition 4 gives some (rather crude) information about the inner structure of the set $A(f)$. Of course, $A(f) \neq \emptyset$, since otherwise $h(\pi_f, A(f)) = \overline{0} \neq \Sigma^d$, a contradiction.

We will now give some more information about $A(f)$. First we consider the case $d = 0$:

Proposition 5. If the assumptions of Theorem 6 are satisfied and if $d = 0$, then the RFDE(f) is point-dissipative. Consequently, the set $A(f)$ is a connected global attractor for the semiflow π_f.

Proof: Let τ be as in the proof of Proposition 2 and $V = V_2$, where V_2 is given by (5). Since $k = d = 0$, it follows that $U = \{0\}$, i.e., $S = C^0$ and hence, noticing that $\Phi(t)\varphi = \varphi\pi_L t$, we have

$$V(\varphi) = \sup_{0 \leq t \leq \tau} [(1+t)||\varphi\pi t||].$$

From the definition of V it follows easily that

$$||\varphi|| \leq V(\varphi) \quad \text{and}$$

$$|V(\varphi) - V(\psi)| \leq (1+\tau)||\varphi-\psi|| \tag{6}$$

for all $\varphi, \psi \in C^0$.

Let $\tilde{f}_n(\varphi) = n^{-1}f(n\varphi)$. As in the proof of Theorem 5 we obtain that for every $\rho \geq 0$, $\sup_{||\varphi|| \leq \rho} ||\tilde{f}_n(\varphi) - L\varphi|| \to 0$ as $n \to \infty$.

Let $\pi_n = \pi_{\tilde{f}_n}$.

A simple estimate implies that

$$\limsup_{t \to 0^+} \frac{1}{t}(V(\varphi\pi_n t) - V(\varphi)) \leq -\frac{1}{1+\tau}||\varphi|| + (1+t)||\tilde{f}_n(\varphi) - L\varphi|| \tag{7}$$

for every $\varphi \in C^0$.

By Theorem 6, A(f) is compact. Hence, in order to show that the RFDE(f) is point-dissipative it suffices to prove that every solution of the RFDE(f) is bounded. Suppose this is not true, and let $t \to x(t)$ be a solution of the RFDE(f) such that $\sup_{t \geq 0} ||x_t|| = \infty$.

There is an n_0 such that for every $n \geq n_0$ and every φ with $||\varphi|| \leq 2(1+\tau)$, $(1+\tau)||\tilde{f}_n(\varphi) - L\varphi|| \leq \frac{1}{2(1+\tau)}$.

There is an $n \geq n_0$ such that $||n^{-1}x_0|| < 1$. Hence, by (6), $V(n^{-1}x_0) < (1+\tau)$. Consequently, there exists a first time $t_1 > 0$ such that $V(n^{-1}x_{t_1}) = (1+\tau)$. Since $t \to n^{-1}x(t)$ is a solution of the RFDE($\tilde{f}_n$), it follows from (6),(7) that for t in a neighborhood of t_1

$$\limsup_{h\to 0^+} \frac{1}{h} (V(n^{-1}x_{t+h}) - V(n^{-1}x_t)) \leq$$

$$\leq -\frac{1}{1+\tau} ||n^{-1}x_t|| + (1+\tau)||\tilde{f}_n(n^{-1}x_t) - L(n^{-1}x_t)|| \leq$$

$$\leq -\frac{1}{1+\tau}(1-\varepsilon) + \frac{1}{2(1+\tau)} < 0.$$

Here $\varepsilon > 0$ is a small number. This implies that $t \to V(n^{-1}x_t)$ is strictly decreasing in a neighborhood of $t = t_1$ and, in particular, that

$$V(n^{-1}x_{t_2}) > V(n^{-1}x_{t_1}) = (1+\tau) \quad \text{for some} \quad t_2 < t_1,$$

a contradiction to our choice of t_1. This contradiction proves that the RFDE(f) is point-dissipative and this, in turn, implies the remaining assertions of the Proposition.

We will now prove, that A(f) has a non-empty unstable manifold, provided $d \geq 1$. Hence in this case, the RFDE(f) is not point-dissipative, and we may expect A(f) to satisfy a saddle-point property, at least generically. However, no proof of the latter conjecture is available.

Proposition 6. If the assumptions of Theorem 6 are satisfied and if $d \geq 1$, then there is a global solution $t \to x(t)$ of the RFDE(f) such that

$$\sup_{t\leq 0} ||x(t)|| < \infty \quad \text{and} \quad \sup_{t\geq 0} ||x(t)|| = \infty.$$

Remark. Hence $x_t \to A(f)$ as $t \to -\infty$ but x_t is unbounded as $t \to +\infty$. Proposition 6 is a special case of Theorem 3.4 in [R3]. The proof is obtained as follows: if the proposition is not true, then every global solution of π_f, bounded on $(-\infty,0]$, is also bounded on $[0,\infty)$. Take a bounded neighborhood N of $A(f)$. It follows that $I^-(N) = I(N) = A(f)$. Using this and the arguments from the proof of Theorem 1, one shows the existence of an isolating block $B \neq \emptyset$ for $A(f)$ such that $B^- = \emptyset$. Hence $h(\pi_f,A(f))$ is the homotopy type of the disjoint union of the set B with a one point-set $\{p\}$. Now an algebraic-topological argument implies that the d-sphere (S^d,s_0), $d \geq 1$, is not homotopy equivalent to such a disjoint union of sets. This is a contradiction and proves the proposition.

Using ideas from the proofs of Theorems 5 and 6 we also obtain the following result:

Theorem 7. *Let* $f\colon C^0 \to \mathbb{R}^m$ *be a locally Lipschitzian mapping. Suppose* φ_0 *is an equilibrium of the RFDE(f), i.e., a constant function such that* $f(\varphi_0) = 0$. *If* f *is Fréchet-differentiable at* φ_0, *and if* 0 *is a hyperbolic equilibrium of the linear RFDE(1),* $L = f'(\varphi_0)$, *with a* d-*dimensional unstable manifold, then* $K = \{0\}$ *is an isolated* π_f-*invariant set,* $h(\pi_f,\{0\})$ *is defined and*

$$h(\pi_f,\{0\}) = \Sigma^d.$$

Sketch of Proof: We may assume w.l.o.g. that $\varphi_0 = 0$. Let $f_\sigma = (1-\sigma)f + \sigma L$, $\pi_\sigma = \pi_{f_\sigma}$, and $K_\sigma = \{0\}$. We claim that there is a closed set $N \subset C^0$ such that N is a strongly π_σ-admissible isolating neighborhood of K_σ, for every $\sigma \in [0,1]$. Assuming this for the moment, we easily see that the map $\sigma \to (\pi_\sigma, K_\sigma)$ is well-defined and $\mathcal{S}$-continuous. Hence Theorem 3 and Corollary 1 imply the result. Now, if our claim is not true, there is a sequence $\sigma_n \in [0,1]$ converging to some $\sigma \in [0,1]$ and a sequence of bounded solutions $t \to x^n(t)$ of $\dot{x} = f_{\sigma_n}(x_t)$ on $(-\infty,\infty)$ such that $0 \neq \sigma_n = \sup_{t\in\mathbb{R}} ||x^n(t)|| \to 0$ and $||x^n(0)|| > \alpha_n - 1$. Let $\tilde{f}_n(\varphi) = (\alpha_n^{-1}) \cdot f_{\sigma_n}(\alpha_n \varphi)$. Then $\tilde{f}_n$ is locally Lipschitzian. Now it is easily seen that $\tilde{f}_n \to L$ uniformly in a bounded neighborhood of zero. Therefore, the arguments from the proof of Theorem 5 lead to a contradiction and complete the proof.

We will now apply our previous results to vector-valued Levin-Nohel equations (cf. Section 3 of these notes). The relevant facts are contained in the following well-known proposition.

Proposition 7 (cf. [H1]). Let $r > 0$ and $b: [-r,0] \to \mathbb{R}$ be a C^2-function such that $b(-r) = 0$, $b'(\Theta) \geq 0$, $b''(\Theta) \geq 0$, for $-r \leq \Theta \leq 0$, and there is a $-r \leq \Theta_0 \leq 0$, such that $b''(\Theta_0) > 0$. Moreover, let $G: \mathbb{R}^m \to \mathbb{R}$ be a C^1-function, and $g = \nabla G$ be locally Lipschitzian. Consider the following RFDE:

$$\dot{x}(t) = -\int_{-r}^{0} b(\Theta) g(x(t+\Theta)) d\Theta. \qquad (8_{b,G})$$

Then the local semiflow $\pi = \pi_{b,G}$ generated by solutions of $(8_{b,G})$ is gradient-like with respect to the following function:

$$V(\varphi) = G(\varphi(0)) + \frac{1}{2}\int_{-r}^{0} b'(\Theta)\left[\int_{\theta}^{0} g(\varphi(s))ds\right]^2 d\Theta.$$

Moreover, every equilibrium φ_0 of $(8_{b,G})$ is constant, $\varphi_0(\Theta) \equiv a$, $\Theta \in [-r,0]$ and $g(a) = 0$.

For the analysis of equilibria of $(8_{b,G})$ we need the following lemma:

Lemma 3. *Let* b *be as in Proposition 7 and let* A *be a symmetric, real-valued* $m \times m$ *matrix. Consider the following linear RFDE*

$$\dot{x}(t) = -\int_{-r}^{0} b(\Theta)Ax(t+\Theta)d\Theta. \qquad (8_{b,A})$$

Then the following properties hold:

1) $\varphi = 0$ *is a hyperbolic equilibrium of* $(8_{b,A})$ *if and only if* A *is nonsingular.*

2) *If* $\varphi = 0$ *is a hyperbolic equilibrium of* $(8_{b,A})$*, then* $h(\pi,\{0\}) = \Sigma^d$*, where* π *is the semiflow generated by* $(8_{b,A})$ *and* $d = d^-(A)$.

Here, $d^-(A)$ *is the total algebraic multiplicity of all negative eigenvalues of* A.

Proof: If $m = 1$, the result is well-known and follows by a simple analysis of the characteristic equation of $(8_{b,A})$. If $m > 1$, let us diagonalize A, using the fact that A is symmetric. We thus obtain that $(8_{b,A})$ is equivalent to a system of m uncoupled one dimensional equations

$$\dot{x}^i = -\lambda_i \int_{-r}^{0} b(\Theta)x^i(t+\Theta)d\Theta, \qquad (9i)$$

where λ_i, $i = 1,\ldots,m$ are the (possibly multiple) eigenvalues of A. Therefore, the unstable manifold of $\varphi = 0$ with respect to $(8_{b,A})$ is easily seen to be $d = d^-(A)$-dimensional. Now Corollary 1 implies the result.

We are now ready to state our main result about equation $(8_{b,G})$.

Theorem 8. Let b, G, and g be as in Proposition 7. Moreover assume the following hypotheses:

1) G is a Morse function, i.e., $G \in C^2(\mathbb{R}^m)$ and whenever $\nabla G(x_0) = 0$, then the Hessian $\left(\frac{\partial^2 G(x_0)}{\partial x_i \partial x_j}\right)_{i,j}$ is nonsingular.

2) There is a symmetric, nonsingular $m \times m$-matrix A_∞ such that

$$\frac{g(x) - A_\infty x}{||x||} \to 0 \quad \text{as} \quad ||x|| \to \infty$$

Then the following statements hold:

i) If x_0 is a zero of g such that $d^-(A_0) \neq d^-(A_\infty)$, where $A_0 = \left(\frac{\partial^2 G(x_0)}{\partial x_i \cdot \partial x_j}\right)_{i,j}$ is the Hessian of G at x_0, then there is another zero x_1 of g and a bounded solution $t \to x(t)$ of $(8_{b,G})$ defined for $t \in (-\infty,\infty)$ such that

either 1° $\lim_{t\to-\infty} x(t) = x_0$ and $\lim_{t\to+\infty} x(t) = x_1$

or 2° $\lim_{t\to+\infty} x(t) = x_0$ and $\lim_{t\to-\infty} x(t) = x_1$.

ii) If $d^-(A_\infty) = 0$, then $(8_{b,G})$ is point-dissipative, hence the union $A(b,G)$ of all global bounded orbits of $(8_{b,G})$ is a connected global attractor.

iii) If $d^-(A_\infty) \geq 1$, then there is a zero $\tilde{x}_0$ of g and a solution $t \to \tilde{x}(t)$ of $(8_{b,G})$ defined for $t \in (-\infty,\infty)$ such that $\lim_{t\to-\infty} x(t) = x_0$ but $\sup_{t\geq 0} ||\tilde{x}(t)|| = \infty$.

The proof of Theorem 8 is an easy consequence of the preceding results.

In the situation of Theorem 8, the union $A(b,G)$ of all full bounded orbits of $\pi = \pi_{b,G}$ is itself bounded, hence compact. Now if $A(b,G)$ is connected, part i) of Theorem 8 is trivial. Hence in order to show the significance of our results it is necessary to prove that $A(b,G)$ is not connected, in general.

In fact, we have the following

Proposition 8. For every $b: [-r,0] \to \mathbb{R}$, $r > 0$ satisfying the assumptions of Prop. 7 and every positive number $c > 0$, there is an analytic function $G: \mathbb{R} \to \mathbb{R}$ such that $g: = G'$ has exactly three zeros $a_1 < a_2 < a_3$, all of them simple, such that $\lim_{|s|\to\infty} \frac{g(s)+cs}{|s|} = 0$, and such that the set $A(b,G)$ is disconnected and consists of the three equilibria $\varphi_i(\Theta) \equiv a_i$, $\Theta \in [-r,0]$, $i = 1,2,3$, and an orbit joining φ_2 with φ_3.

Proof: Choose $\tilde{a}_1 = 0$. Let $f(s) = -cs$. Then, there is a unique $\lambda > 0$ such that $x(t) = e^{\lambda t}$, $t \in \mathbb{R}$, is a solution of $(8_{b,F})$ where $F(x) = \int_0^x f(s)ds$. Let $t_0 > 0$ be arbitrary and let $y(h)$, $h \in [0,r]$, be defined as

$$y(h) = x(t_0) - \int_0^h \left(\int_{-r}^{-s} b(\Theta)f(x(s+t_0+\Theta))d\Theta\right)ds.$$

Hence, there is a $0 < h_1 < r$ such that $y(h) > 0$ for $h \in [0,h_1]$, hence $y(h_1) > x(t_0)$. Define a continuous function $\tilde{f}$ such that $\tilde{f} = f$ on $(-\infty,x(t_0)]$, $\tilde{f}$ is affine on $[x(t_0),y(h_1)]$, $\tilde{f}(y(h_1)) = 0$ and $\tilde{f}$ is affine on $[y(h_1),\infty)$ with negative slope (see Fig. 5) equal $-c$.

If $\varphi(\Theta) = x(t_0+\Theta)$, $\Theta \in [-r,0]$, let $\tilde{y}$ be the solution of $(8_{b,\tilde{F}})$ through φ, where $\tilde{F}(x) = \int_0^x \tilde{f}(s)ds$. Then obviously $\tilde{y}(h) > y(h)$ for $h \in (0,h_1]$. If we define $\tilde{x}(t) = x(t)$, $t \leq t_0$, $\tilde{x}(t) = \tilde{y}(t-t_0)$, $t > t_0$, then $\tilde{x}$ is a solution of $(8_{b,\tilde{F}})$ on $\mathbb{R}$. Moreover, $\tilde{x}(t) \to \infty$ as $t \to \infty$, for otherwise $\tilde{x}(t)$ would go to a zero β of $\tilde{f}$, $\beta > y(h_1)$, a contradiction. Consequently, there is an $s_1 > t_0$ such that $\tilde{x}(s_1+\Theta) > y(h_1)$ for $\Theta \in [-r,0]$. Now perturb $\tilde{f}$ a little on an interval $[x(t_0)-\varepsilon,y(h_1)]$, where $\varepsilon > 0$ is a small number, to obtain a C^1-function $\tilde{g}$ which has exactly three simple zeros $\tilde{a}_1 = 0 < \tilde{a}_2 < \tilde{a}_3 = y(h_1)$ (see Fig. 6).

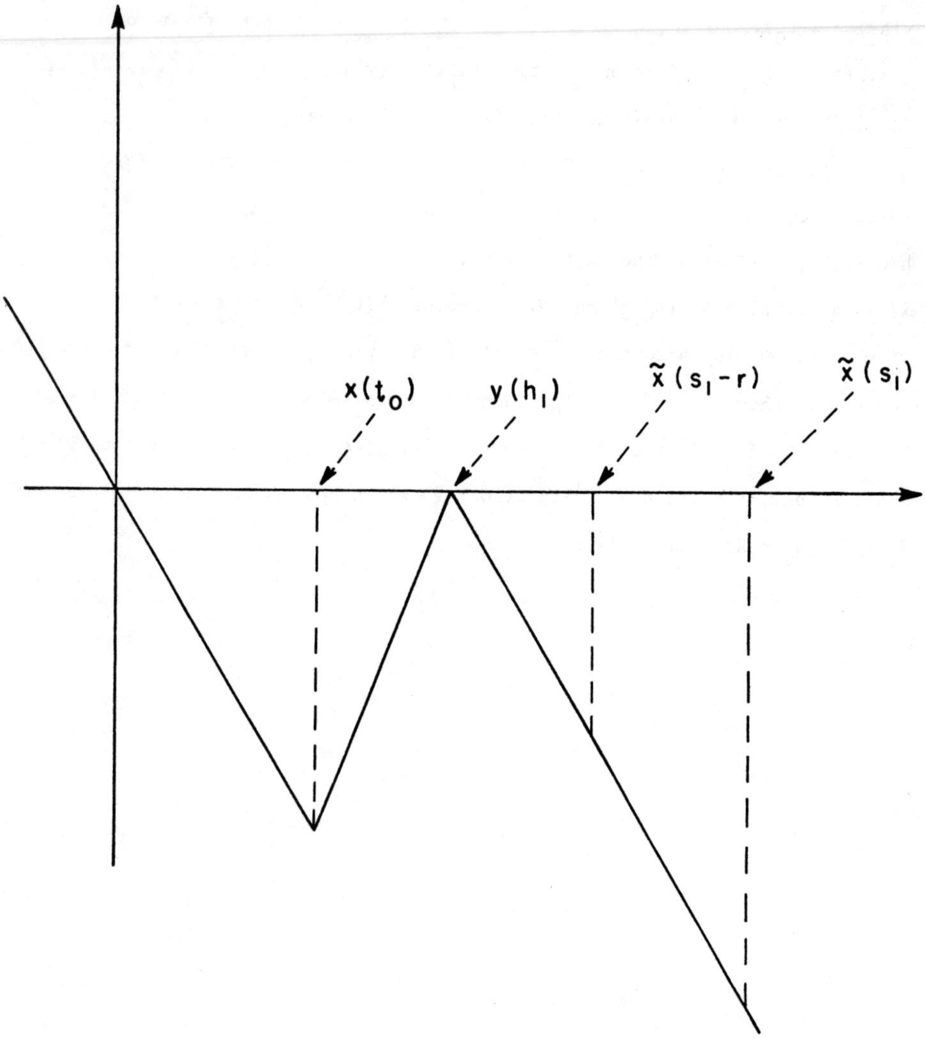

Figure 5

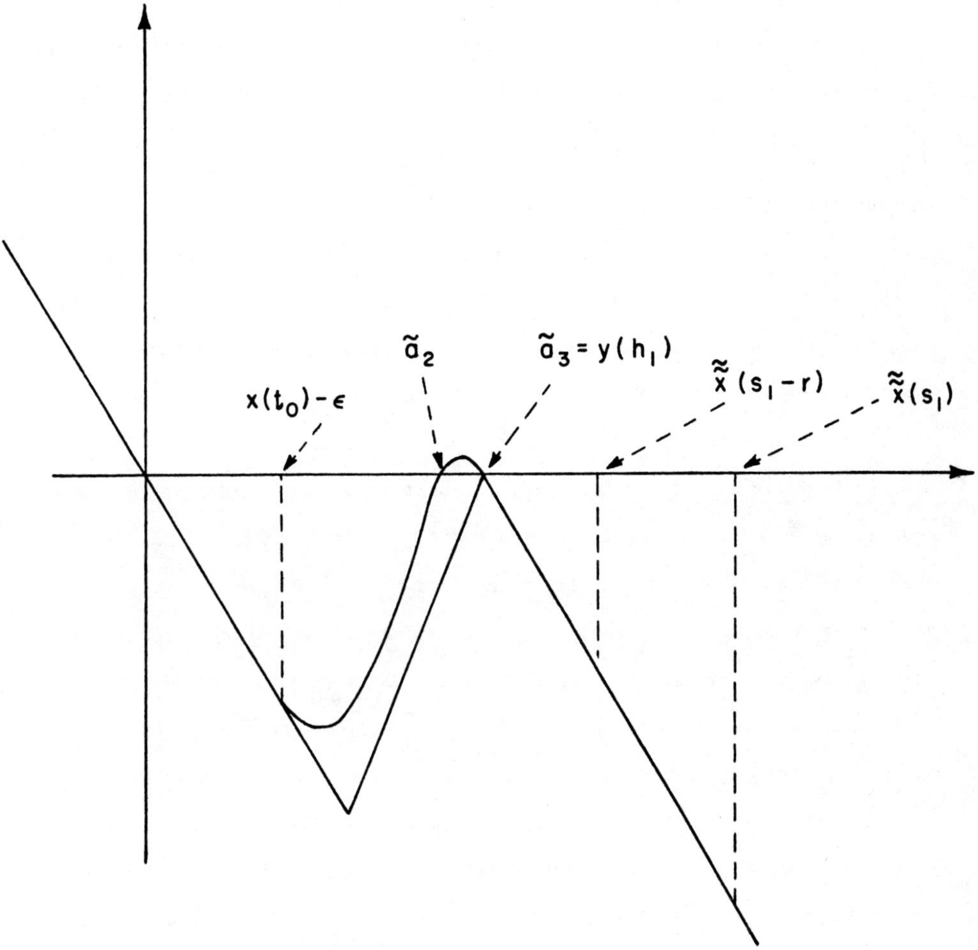

$\tilde{a}_2$
$\tilde{a}_3 = y(h_1)$
$x(t_0) - \epsilon$
$\tilde{\tilde{x}}(s_1 - r)$
$\tilde{\tilde{x}}(s_1)$

Figure 6

Let $\tilde{G}(x) = \int_0^x \tilde{g}(s)ds$. If the perturbation is small, then the unique solution $t \to \tilde{\tilde{x}}(t)$ of $(8_{b,\tilde{G}})$ which emanates from $\tilde{a}_1 = 0$ (i.e. $\tilde{\tilde{x}}(t) \to \tilde{a}_1$ as $t \to -\infty$) and staying to the right of $\tilde{a}_1$ is such that $\tilde{\tilde{x}}(s_1+\Theta) > y(h_1)$ for $\Theta \in [-r,0]$. Hence $\tilde{\tilde{x}}(t) \to \infty$ as $t \to \infty$ and this implies that there is no bounded orbit of $(8_{b,\tilde{G}})$ emanating from $\tilde{a}_1 = 0$.

Furthermore the unique orbit emanating from $\tilde{a}_3 = y(t_1)$ and staying to the left of $\tilde{a}_3$ must run to $\tilde{a}_2$. In fact it cannot hit $\tilde{a}_1$ as is easily seen by examining the Liapunov function V of Proposition 7: as $t \to \tilde{\tilde{x}}(t)$ emanates from $\tilde{a}_1$ and hits $\tilde{a}_3$, it follows that $\tilde{G}(\tilde{a}_3) < \tilde{G}(\tilde{a}_1)$. Hence there is no orbit emanating from a_3 and hitting a_1, because otherwise $\tilde{G}(\tilde{a}_3) > \tilde{G}(\tilde{a}_1)$. It follows that the set $A(b,\tilde{G})$ defined above consists of the three equilibria $\tilde{\varphi}_i = \tilde{a}_i$, $i = 1,2,3$ and an orbit running from $\tilde{a}_3$ to $\tilde{a}_1$. Hence the Proposition is proved except for the fact that $\tilde{g}$ is not analytic. Now one can approximate $\tilde{g}$ by analytic functions g using Whitney's Lemma. Alternatively, one may use the proof of Lemma 2.5 in [HR] to conclude that if g is sufficiently close to $\tilde{g}$ and $G(x) = \int_0^x g(s)ds$, then G, g and $A(b,G)$ satisfy all the statements of our Proposition. The proof is complete.

We can compute the index $h(\pi_{b,G}, A(b,G))$ as follows: Let $g_\sigma(s) = -\sigma\cdot c\cdot s + (1-\sigma)g(s)$, $\sigma \in [0,1]$, and $G_\sigma(x) = \int_0^x g_\sigma(s)ds$. Then it easily follows that the map $\alpha\colon \sigma \to (\pi_{b,G_\sigma}, A(b,G_\sigma))$ is well-defined and $\mathcal{S}$-continuous. Consequently, by Theorem 3 and Lemma 3

$$h(\pi_{b,G}, A(b,G)) = h(\pi_{b,G_1}, \{0\}) = \Sigma^1.$$

Moreover, by Theorem 7,

$$h(\pi_{b,G},\{\varphi_i\}) = \begin{cases} \Sigma^1 & \text{if } i = 1,3 \\ \Sigma^0 & \text{if } i = 2. \end{cases}$$

This illustrates very clearly the concept of irreducibility: there is no heteroclinic orbit running from or to φ_1. By Proposition 4, this is only possible if the index of $(\pi_{b,G},\{\varphi_1\})$ is either $\overline{0}$ or equal to the index of $(\pi_{b,G},A(b,G))$, and this is indeed the case. By the same token, since $h(\pi_{b,G},\{\varphi_2\}) = \Sigma^0 \neq h(\pi_{b,G},A(b,G))$, it follows from Proposition 4 that φ_2 is the "target" or the "source" of a heteroclinic orbit, the former being the case here.

Note that Proposition 3 gives no criterion to detect heteroclinic orbits emanating from or tending to an equilibrium φ_i, if the index of this equilibrium is $\overline{0}$ or equal $h(\pi_{b,G},A(b,G))$. In fact, in our example, $\{\varphi_1\}$ is isolated in $A(b,G)$ and $\{\varphi_3\}$ is the source of a heteroclinic orbit, although both equilibria have the same index equal $h(\pi_{b,G},A(b,G)) = \Sigma^1$.

Concluding Remarks.

In this Appendix, we have only presented the simplest aspects of the homotopy index theory on noncompact spaces. In particular, we entirely omitted the discussion of the Morse index as a category (see [Co] and [R2]). In many cases, invariant sets K admit a so-called Morse decomposition. Classical examples include finite sets of equilibria in K. A question arises as to the existence of heteroclinic orbits connecting such equilibria. We discussed this question above in a very simple setting, but much more can be said leading to the notion of index triples, the connection-index and generalized Morse inequalities ([Co], [R2], [RZ]). Recently, J. Mallet-Paret (oral communication) introduced an interesting gradient-like structure on a class of scalar delay equations. This structure is induced by a Liapunov function which is analogous to the so-called lap-number of Matano for parabolic equations. Analyzing the Morse decomposition thus obtained and applying arguments from the Morse-index theory, the author is able to prove the existence of special "periodic" solutions of a singular perturbation problem $\varepsilon\dot{x} = f(x(t), x(t-1))$.

In the applications of the index theory in this section, only the so-called non-resonance case was considered. In particular, the equilibrium 0 of f was assumed to be hyperbolic. If this assumption is dropped, then there is a local center manifold at zero, which contains all small invariant sets of π_f. One can then show that the index of every such small isolated invariant set with respect to π_f is a "product" of the index of the same set relative to the center manifold with Σ^m, where m is the dimension of the unstable manifold of 0. Using this product formula, one can obtain information about $A(f)$ in this resonance case. For an application of such arguments to PDEs, see [R6] and [R7].

References

[CE] C. C. Conley and R. Easton, Isolated invariant sets and isolating blocks, TAMS 158 (1971), 35-61.

[ChH] S. N. Chow and J. K. Hale, Methods of Bifurcation Theory, Springer-Verlag, 1982.

[Co] C. C. Conley, Isolated invariant sets and the Morse index, CBMS, No. 38, Providence, R. I., 1978.

[H1] J. K. Hale, Theory of Functional Differential Equations, Springer-Verlag, 1977.

[H2] J. K. Hale, Topics in Dynamic Bifurcation Theory, CBMS Lecture Notes, Vol. 47, Am. Math. Soc., Providence, R. I., 1981.

[HR] J. K. Hale and K. P. Rybakowski, On a gradient-like integro-differential equation, Proc. Roy. Soc. Edinburgh, 92A (1982), 77-85.

[R1] K. P. Rybakowski, On the homotopy index for infinite-dimensional semi-flows, TAMS 269 (1982), 351-383.

[R2] K. P. Rybakowski, The Morse index, repeller-attractor pairs and the connection index for semiflows on noncompact spaces, JDE 47 (1983), 66 98.

[R3] K. P. Rybakowski, On the Morse index for infinite-dimensional semiflows, in: Dynamical Systems II (Bednarek/Cesari, eds.), Academic Press, 1982.

[R4] K. P. Rybakowski, Trajectories joining critical points of nonlinear parabolic and hyperbolic partial differential equations, JDE, to appear.

[R5] K. P. Rybakowski, Irreducible invariant sets and asymptotically linear functional differential equations, Boll. Unione Mat. Ital., to appear.

[R6] K. P. Rybakowski, An index-product formula for the study of elliptic resonance problems, submitted for publication.

[R7] K. P. Rybakowski, Nontrivial solutions of elliptic boundary value problems with resonance at zero, submitted for publication.

[RZ] K. P. Rybakowski and E. Zehnder, A Morse equation in Conley's index theory for semiflows on metric spaces, Ergodic Theory and Dyn. Systems, to appear.

[PS] R. Palais and S. Smale, Morse theory on Hilbert manifolds, Bull. Amer. Math. Soc. 70 (1964), 165-171.

[WY] F. W. Wilson and J. A. Yorke, Lyapunov functions and isolating blocks, JDE 13(1973), 106-123.

Index

Index

Index